H. Dupuis   G. Zerlett

The Effects
of Whole-Body Vibration

Foreword by H. E. von Gierke

With 54 Figures

Springer-Verlag Berlin Heidelberg GmbH

Professor Dr. Heinrich Dupuis
Institute of Occupational and Social Medicine
Director of the Ergonomics Working Group
Johannes Gutenberg University, Mainz,
Federal Republic of Germany

Dr. Georg Zerlett
Main Department of Occupational Medicine
Director of the Department of Industrial Medical Services
Rheinische Braunkohlenwerke AG, Cologne,
Federal Republic of Germany

ISBN 978-3-540-16584-2   ISBN 978-3-642-71245-6 (eBook)
DOI 10.1007/978-3-642-71245-6

Library of Congress Cataloging-in-Publication Data
Dupuis, Heinrich. The effects of whole-body vibration.
Translation of Beanspruchung des Menschen durch mechanische Schwingungen.
Bibliography: p.
1. Vibration––Physiological effect. I. Zerlett, Georg, 1927–
II. Title. QP82.2.V5D8713 1986 612′.01445 86-6772

© Springer-Verlag Berlin Heidelberg 1986
Originally published by Springer-Verlag Berlin Heidelberg New York Tokyo in 1986

2119/3145-543210

ACKNOWLEDGEMENTS

The authors would especially like to express their gratitude to:
Drs. W. Christ, L. Vogt, and E. Hartung for their support and
cooperation; medical technicians, Mrs. W. Gilsdorf and Mrs. U.
Meyersen, for their help in building up the technical library on vibration; Mrs. S. L. Faatz for the translation; Professor H. von Gierke
and Dr. J. Sandover for correcting the specialized terminology
and technical sections of the translation; and all other colleagues
who have contributed to the preparation of this book.

Foreword to the English Edition

It is with great pleasure and satisfaction that I welcome the English edition of *The Effects of Whole-Body Vibration* by Drs. Dupuis and Zerlett. Although it might appear that human vibration exposure is a relatively limited specialized field, it is impressive to see how many investigations have been conducted in various countries over the last 50 years and to note the different reasons for which they were undertaken. The results are that they can be found not only in many languages but also in many different scientific and technical fields from architectural, railroad and automobile magazines to occupational medicine, aerospace medicine, human factors, engineering mechanics, and many other journals. The different national languages do not prevent communication between researchers and users of the data as much as do the different technical languages and different goals of past investigations in this field.

A critical monograph, summarizing the field, with a senior author who actively contributed to this research over the last quarter century, is therefore of outstanding value. It should contribute to the cross-fertilization of the fields involved and provide the basis for interdisciplinary cross-fertilization and understanding. The English translation, no doubt, will contribute to international communication and collaboration by bringing many not easily accessible papers, primarily German and European contributions, to the attention of English-speaking colleagues. That the reporting has gone through the critical − but naturally sometimes unavoidably subjective − interpretation of the authors is no drawback and might stimulate readers to study the original references.

Who might benefit from reading the book or using it as a reference? With elementary introductions to all subfields such as physiology, medicine, physics and mechanics, it should be useful to researchers and practitioners in fields such as occupational and orthopedic medicine; health professionals and industrial hygiene engineers; design engineers in the aviation, automotive, ship building, agricultural machinery, personal protection industry; human engineers; ergonomists and many other disciplines. Although the relationship between the effects of whole-body vibration, impact exposure and crash protection research is outside the scope of this book, it is obvious that many chapters of the book can make valuable contributions to such neighboring areas of biodynamics.

It is hoped that this monograph will contribute to one area of particular interest to this writer: national and international vibration exposure standards with respect to health, increased working efficiency and improved comfort. The communication opened up by this book should prove very valuable.

Yellow Springs, Ohio, January 1986

Henning E. von Gierke

Ph. D., Engineering

Director, Biodynamics and Bioengineering Division
Armstrong Aerospace Medical Research Laboratory

Clinical Professor, Department of Community Medicine
Wright State University

Chairman, Subcommittee "Human Exposure to Vibration and Shock" ISO/TC 108/SC4

Preface to the German Edition

The observations made by Paracelsus concerning the dose-effect relationship of poison are generally just as applicable to health-threatening vibration. With regard to kind, intensity, and duration of the vibration, the "dose" is decisive as to whether the consequences are detrimental, unmeaningful, or tolerable with respect to health. This law of nature determines the tasks and goals of those whose aim is to safeguard health. Researchers worldwide have been occupying themselves with this many-faceted question for some time: how mechanical vibration affects the human organism and at what point damage occurs. If prevention in occupational medicine is to succeed, it is most important that the gaps in our present knowledge be closed, for if technical preventive measures are to be effective and preventive means in occupational medicine successful, they must be based on reliable and complete findings.

Whenever many independent researchers have worked in a given field for a long period, a comprehensive intermediate assessment is appropriate to evaluate the level achieved and the direction the research is going. In the area of hand-arm vibration, this evaluation has already been carried out, and the response aroused by this particular research report had led to a demand for a comparable evaluation of the research results on whole-body vibration.

This report presents clearly and exhaustively the current status of international knowledge, as well as the questions that remain to be answered. The authors, themselves recognized specialists active in the research in question, have made a significant contribution to the dissemination of existing knowledge and to the clinical orientation of ongoing research.

Since the quality of the report and the international nature of the subject matter guarantees worldwide interest, it is assumed that publication in English will meet with wide approval.

This research project was financially supported by the Hauptverband der gewerblichen Berufsgenossenschaften e.V. (Central Association of the Industrial Injuries Insurance Institutes).

Bonn, May 1984

von Hassel	R. Hopf	Dr. Waterman
Chairman	Deputy Chairman	General Manager
Board of Directors	Board of Directors	Central Association

Table of Contents

1. Introduction

With the advent of modern technology, new environments have also been created that have a strong influence on daily life. Noise and mechanical vibration belong to the new environmental factors that affect both our work and leisure time. People can be affected in different ways by mechanical vibration, also known as shock or oscillation: by contact with vibrating machinery, when using means of transportation (as passenger or driver), and even when occupying buildings.

In former times, animals were used for transport and — after the invention of the wheel — horse-drawn carriages. Only a relatively small proportion of the population was affected by vibration from these means of transport: e.g., couriers, officers, or the cavalry. In 1718, Ramazzini, the founder of systematic occupational medicine, observed in his book *Studies on the Diseases of Artists and Artisans* that, of 42 occupations, only "horse-breaking" resulted in stress and strain from mechanical vibration. He describes the mechanical effects as follows:

... the whole intestinal tract was shaken by the force and dislocated from its natural position ... Likewise, constant riding caused tearing of the breast vessels ...

Today people traveling by land, sea, and air — whether the driver/pilot/captain, passenger, or the sick and injured being transported — are all exposed to mechanical vibration. For example, truck drivers, railroad conductors, the crew or passengers of airplanes, helicopters, or ships are all involved. It is roughly estimated that every day several million employees and passengers of public, private, and industrial means of transport are exposed to mechanical vibration in the Federal Republic of Germany. Furthermore, drivers of heavy equipment, such as earth-moving equipment, tractors, street-cleaning machines, and land or forestry tractors, are also affected.

However, stationary machines (e.g., generators, turbines, sledgehammers, presses, engine test machines) can also produce vibrations that can have an effect on the operator. Finally, machines and vehicles (e.g., undergrounds or subways) can also produce vibrations in buildings and parts of buildings, which in turn are transmitted to people.

In the last 20 years, many research reports and other documents have been published on the effects of vibration stress that deal with different types of vibration and intensity ranges. Some papers (Coermann 1965; von Gierke 1968; Dupuis 1969; Seidel 1975; Heide 1977; Griffin and Lewis 1978; Rublack 1978;

Lewis and Griffin 1978) also try to present comprehensive surveys of current knowledge on the effects of "whole-body vibration." Since important scientific works on this theme have been completed in recent years, it seems meaningful and necessary to present a comprehensive review of the status of physiological findings in occupational medicine, with regard to the effects of whole-body vibration, as analyzed from the international literature.[1] For the following reasons, this overview is especially urgent and of present interest.

– Experience shows that whole-body vibration, dependent on the kind, intensity and duration, has various effects, which can result in disturbances of well being, perception of pain, physiological reactions, and decrease in performance.

– Because of "atypical (nonspecific) complaints" and the difficulty in proving that there is an occupation-specific origin, the illnesses resulting from low-frequency whole-body vibration so far have not been included in the list of occupational diseases compiled by the International Labor Office (ILO). In Para. 26 of the revised list of occupational diseases in ILO Agreement No. 121 (ILO 1980), it has been established, however, "that the experts recognize the importance of such illnesses in tractor and trailor drivers and that they are of the opinion that, in this case, there is a serious preventive problem."

– Furthermore, it has been stated in ILO Agreement 148 (1977) that: measures have to be taken to protect employees from vibration; the responsible authorities have to establish criteria to determine the danger; when necessary, the exposure limits must be defined by means of these criteria. Supervision of employees exposed to occupational hazards as a result of vibration at their places of work must also include a medical examination before the beginning of this particular job, as well as regular check-ups later on.

– With the goal of developing "occupational health guidelines for whole-body vibration," preliminary work is currently being done to establish the criteria for preventive medical examinations in occupational medicine.

– Furthermore, the Central Association of the Industrial Injuries Insurance Institutes is also working on a draft outline of regulations for accident prevention, "Vibration."

– With regard to the question of compensation for work-related illnesses, law suits frequently occur that require a decision as to whether a certain injury to the spine, for instance, can be traced back with a high degree of probability to previous strong vibration stress and whether it can be recognized under the corresponding requirements. Therefore, expert testimony must take into consideration the current knowledge in occupational medicine.

Therefore, it is the goal of this report to demonstrate on the basis of a comprehensive study of the literature the present level of knowledge in occupational medicine with regard to the effects of whole-body vibration. In this way, assistance may be provided to the Ministry of Labor and Social Affairs and to the preparation of occupational health regulations regarding preventive measures;

---

[1] In this connection, the specific effects of mechanical vibration on the hand-arm system will be disregarded, as a comprehensive analysis of the literature has recently been completed (Dupuis 1982)

legal experts and courts can also be helped in legal decisions concerning occu-pational medicine.

The other goal is to obtain scientific answers to this question: How can pre-ventive measures reduce the risk of illness caused by whole-body vibration? Further initiatives may be prompted to introduce Agreement No. 148 of the ILO (1977), with regard to whole-body vibration, into German law.

2. Terminology and Definitions

In order to eliminate as much misunderstanding as possible in the evaluation of various forms of vibration stress, it is necessary to define the technical physical concepts used (Table 1). In addition, the special medical terms are explained at the end of the book (p. 158).

When discussing stress resulting from whole-body vibration, the direction of the vibration, x, y and z, refers to a coordinate system relating to the human body (Fig. 1). This coordinate system was established nationally in 1979 (VDI 2057) and internationally in 1978 (ISO 2631).

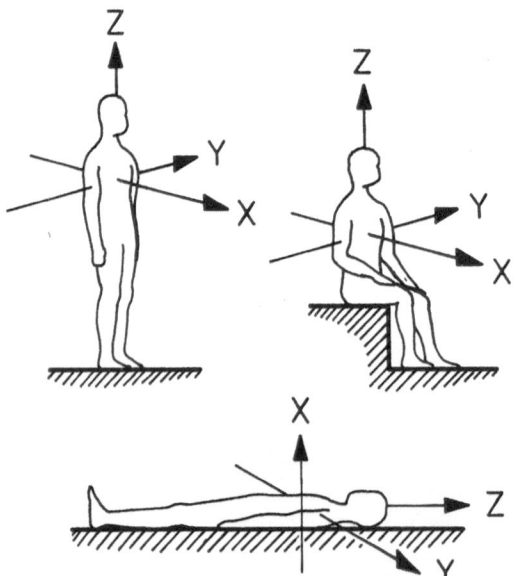

Fig. 1. Drawings showing the direction of the vibration, x, y and z, which refers to a coordinate system relating to the human body (VDI 2057, p. 1; ISO 2631)

Table 1. Vibration terminology[a]

Term	Symbol	Dimension	Definition
Amplitude		m, m/s, m/s²	Momentary maximum value of a quantity
Broad-band vibration			Vibration with frequency content of more than one third octave (terz)
Crest factor			Quotient from peak-value and r.m.s. value
Damping			Transformation of vibration energy to warmness
Duration of period	T	s	Time of a single vibration (period of vibration)
Evaluation filter			In regard to ISO 2631 and VDI 2057, an active or passive electrical network. The frequency-dependent transmission factor of that considers the frequency-dependence of the vibration effects (see: frequency-weighted vibration and K-value)
Frequency	f	1/s	Number of periods per time and reciprocal value of duration of period T
Frequency-weighted acceleration	a_w	m/s²	Frequency-weighted acceleration according to ISO 2631 and VDI 2057 (see: K-value) $a_{wx} = KX/28$ $a_{wy} = KY/28$ $a_{wz} = KZ/20$
Harmonic vibration			Synonymous term for sinusoidal vibration
K-value	K		(German: *Bewertete Schwingstärke K,* earlier: *Wahrnehmungsstärke K*), measure of relation between objective (measurable) vibration stress on the one hand and subjective (not measurable) perception, biodynamic behavior, and certain physiological changes on the other, according to German VDI recommendation 2057 (see also: frequency-weighted acceleration). Additional letter L pertains to the reclining posture, B to exposure in buildings, and X, Y and Z to the vibration direction according to Fig. 1 (see p. 4)
Natural frequency	f_0	1/s	When a vibration system is excited to vibration by a solitary impulse, it will move in its natural frequency (resonance frequency)
Mechanical impedance	Z	N · s/m	Complex ratio of force to velocity where the force and velocity may be taken at the same or different points in the same mechanical system during simple harmonic and steady-state motion
Mechanical vibration			Variation with time of the magnitude of a quantity which is descriptive of the motion or position of a mechanical system, when the magnitude is alternately greater and smaller than some average value or reference; synonymous term for shaking, oscillation
Oscillation (motion)			Synonymous term for mechanical vibration, shaking

Table 1. (continued)

Term	Symbol	Dimension	Definition
Peak value			Maximum value of a quantity during a given interval
Periodic vibration			Periodic quantity whose values recur for certain equal increments of the independent variable
Random vibration			Vibration whose magnitude cannot be precisely predicted for any given instant of time
Resonance			Resonance of a system in forced oscillation exists when any change, however small, in the frequency of excitation causes a decrease in a response of the system
Root-mean-square value of acceleration	a_{rms} or a_{eff}	m/s^2	Square root of the average of the squared acceleration values along averaging in T
Rotational vibration			Circular motion about a rotation axis roll: about x-axis pitch: about y-axis yaw: about z-axis
Shaking			Synonymous term for mechanical vibration, oscillation
Sinusoidal vibration			Vibration with periodic quantity of a sinusoidal or cosinusoidal function
Small-band vibration			Vibration with frequency content of less than one-third octave (terz)
Spring rate	c	N/m	Quotient from spring force and spring excursion (displacement)
Transient vibration			Vibration of a system other than steady-state or random, short-term motion with shock-type character
Translational vibration			Rectilinear motion
Transmission factor	V		Quotient from response amplitude to exciting amplitude
Vibration acceleration	a	m/s^2	Vector quantity that specifies the time-derivative of velocity (change of vibration velocity per time)
Vibration displacement		m	Vector quantity that specifies the change of position of a body or particle, with respect to a reference frame
Vibration force	F	N	Force transmitted at the place of entrance into the human body
Vibration strain			Effects of vibration stress (load) on man (response), as discomfort, decreased proficiency, injury to health
Vibration stress			Vibration from outside act on man, to be measured at place of entrance
Vibration velocity	v	m/s	Vector quantity that specifies the time-derivative of displacement (change of vibration displacement per time)

[a] With reference to: ISO 2041 (1975), ISO 5805 (1981), ISO 5982 (1981), DIN 1311 (1974), VDI 2062 (1976), VDI 2057 Bl. 1 (1983), Bosch (1976), Harris and Crede (1961), Klotter (1978), Dupuis (1975a, 1981a, 1981b)

3. Stress-Strain Concept in Whole-Body Vibration

The strain-stress concept defines stress. First of all, stress refers to vibration exposure (kind, intensity, and duration of vibration) and the additional stress factors that act upon man when exposed. Strain, on the other hand, means the effects of such stress, that is, the reaction (biodynamic, psychological, physiological, damaging) of the human being, taking into consideration the individual endogenous conditions (see Table 2).

To evaluate the risk of strain from whole-body vibration, it is essential to look into both the precise work case history and – in individual evaluations – the clinical case history of given endogenous requirements. The contributing

Table 2. Schematic representation of stress and strain from whole-body vibration

Work-related stress (exogenous stress)			Individual endogenous preconditions
Work history			*Clinical case history*
Physical stress (vibration characteristics	Other contributory factors	Duration of exposure	
Amplitudes	Body posture	Daily	Sex
Frequencies	Seat belts	Annually	Age
Shocks	Active and passive	During profes-	Age at beginning of job
Vibration directions	support	sional life	Constitution
		Break routine	Disposition
			Endogenous condition of spinal column

Strain from whole-body vibration

Acute effects
Subjective discomfort
Pain perception
Biomechanical reactions
Physiological reactions
Decrease in performance

Chronic effects
Injury to health

factors in stress and strain are arranged schematically in the graph shown in Table 2.

In work-related stress (exogenous stress), which in individual cases must be understood by means of the work case history, the physical vibration stress from vehicles or machinery, other contributing physical stress factors, and the duration of exposure all have to be taken into consideration. It is understandable that the kind and amount of these stress factors, which depend upon the kind of vehicle as well as on the kind of machine, can be very different from the motion and work conditions.

For non-work-related stress, such as the vibration stress that occurs in passengers or apartment tenants, similar studies need to be carried out concerning the kind and intensity of stress.

The German *Berufsgenossenschaftliches Institut für Arbeitssicherheit* (BIA), in St. Augustin (Occupational Safety Institute of Industrial Injuries Insurance Institutes), is collecting data on vibration stress in various places of work (vibration data bank). A comprehensive survey on stress from whole-body vibration is given in the section called "Work-Related Stress from Whole-Body Vibration" in this report (see p. 9).

In addition to exogenous stress, the endogenous prerequisites of the person in question play an important role regarding the risk of strain. This can be explained by using the spinal column as an example. As will be shown in the section on "Diseases of the Spinal Column" (see p. 91), the spinal column is especially endangered by vibration (Dupuis 1980b). The age of the worker when first exposed to vibration at work is important as long as the growth of the spinal column is not yet complete. With advancing age, the strength and elasticity of the spinal column as a whole are reduced, which results in the reduced capacity of the spinal column to bear stress from mechanical vibration. With regard to differences in constitution and disposition, we have to assume, among other factors, that well-developed back, thoracic and abdominal musculature has a positive function as a "muscle corset." It provides support for the spinal column under vibration stress. Furthermore, it can also be assumed that the endogenous condition of the spinal column can undoubtedly have a distinct influence on the capacity to bear stress.

The effects of exogenous stress, combined with the endogenous factors, are decisive with regard to the amount of strain that can be expected during whole-body vibration. This strain can be subdivided into acute effects (disturbances of well-being, pain, biomechanical, vegetative and other physiological reactions, and reduction in performance) and chronic effects (injury to health). This is essentially the content of this report.

4. Work-Related Stress from Whole-Body Vibration

During the last 10 years, many field experiments have been carried out in various places of work. Vibration has been measured in buildings, in vehicles, and on ships. These data are currently being compiled (Dupuis and Hartung 1982) to be fed into the vibration data bank at the BIA.

Since 1963, the K value, a dimensionless quantity which also accounts for the frequency dependency of the effects on man, has been used in the Federal Republic of Germany to characterize vibration stress.

This K value was initially designated as "perception quantity," but since 1981 it has been called (frequency-) "weighted vibration quantity" because it not only takes into consideration the frequency dependency of subjective perception but also the objectively measurable reaction of the human body. Conversion of the K value into the internationally established value, a_w (frequency-weighted acceleration) − ISO 2631, is possible for the three directions of vibration, x, y and z, of a standing or sitting person in accordance with the following formulas:

$$a_{wz}\,[\mathrm{m/s^2}] = \frac{KZ}{20\,(\mathrm{m/s^2})^{-1}}$$

$$a_{wx,\,y}\,[\mathrm{m/s^2}] = \frac{KX}{28\,(\mathrm{m/s^2})^{-1}} = \frac{KY}{28\,(\mathrm{m/s^2})^{-1}}$$

(For the reclining person, conversion of the KL value to the a_w value is not possible, since there is no ISO convention for this type of effect.)

Table 3 presents an appropriate survey of the results from measurement of the vibration stress (for seated persons in moving vehicles, in moving work equipment, on ships, and in helicopters). The vibration values given there are to be understood in terms of orientation aids, which are designed to illustrate the magnitude of the assessed vibration in the z-direction at various places of work. Only data measured at representative places of work were selected from the literature. However, data on extreme conditions, such as test drives, are not included in the table.

The data in Table 3 are taken from publications such as the following: Bartels et al. (1981); Bez (1980); Brand and Schnauber (1980); Dupuis and Hartung (1972); Dupuis and Hartung (1973), Dupuis (1980a); Frenking (1980); Hilfert et al. (1981); Schmidt (1980); Köhne (1982); Köhne et al. (1981); Dupuis et al. (1982b).

Table 3. Weighted K value in vehicles, helicopters, and buildings, as well as on moving work equipment and ships

Vehicle/place of work	Frequency-weighted acceleration $a_{wz\,rms}$ (m/s^2)	K value[a] (KZ)
Road vehicles with suspension		
Automobiles	0.2 −0.75	4−15
Buses	0.4 −0.8	8−16
Ambulances	−	4−25 (KXL)
Trucks (road)	0.2 −0.9	4−18
Trucks (building sites)	0.7 −1.4	14−28
Work equipment without suspension		
Forging equipment	0.25−0.75	5−15
Agricultural tractors	0.4 −1.25	8−25
Forklifts	0.4 −2.0	8−40
Excavators	0.3 −1.1	6−22
Crawler levelers	0.3 −1.3	6−26
Rollers	0.3 −1.7	6−34
Graders	0.95−1.6	19−32
Wheel dozers	0.6 −2.2	12−44
Wheel loaders	0.5 −2.4	10−48
Track vehicles		
Track cranes	0.05−0.25	1− 5
Bridge cranes	0.1 −0.8	2−16
Switch locomotives	0.2 −0.7	4−14
Freight and passenger locomotives	0.3 −0.6	6−12
Military vehicles		
Trucks, old generation	1.3 −4.0	26−80
Trucks, new generation	0.45−1.65	9−33
Tanks	1.5 −3.5	30−70
Ships		
Crew quarters	0.5 −0.7	10−14
Bridge	0.2 −0.35	4− 7
Other quarters	< 0.2	< 4
Helicopters	0.1 −1.55	2−31

[a] Orientation values not valid for extreme conditions

Table 3 shows that vibration stress in road vehicles is relatively low. The K value data in an ambulance refers to the reclining patient in the vertical x direction. In earth-moving equipment, such as loaders, crawler levelers, graders, wheel dozers, excavators, and compactor rollers, relatively high vibration values can generally be expected in the z-direction. In bridge and track cranes, the vibration stress can usually be considered as low. In contrast, one can find stress in military vehicles, as they are submitted to special operating conditions.

On ships, the vibration stress is comparatively low. However, the fact that the crew is exposed to vibration for 24 h/day must be taken into consideration.

In the horizontal vibration directions, x and z, the vibration stress is generally less in moving vehicles and moving work machinery than the values given in Table 3 for the vertical direction. Vibration intensity in the x and y directions usually amount to only 50% − 70% of the value in the z-direction.

As reported by Schnauber and Weigelt (1981), in standing jobs, such as when using a sledgehammer, operating a press, grinding or form machines, and on diesel locomotives, the vibration stress mostly ranges below $a_w = 0.5$ m/s². Vibration in buildings can reach an assessed a_w value of $0.005 - 0.031$ m/s².

The daily vibration exposure time in the aforementioned places of work is usually less than 8 h (daily work shift). (Exception: large sea-going ships with around-the-clock 24-h exposure.)

5. Acute Effects of Mechanical Vibration

5.1 Biological Prevention and Control Mechanisms Against Mechanical Vibration

We know that man possesses a highly developed ability to adjust to different environmental conditions. For this function there are physiological regulatory systems, which release reactions in him for protection against such influences. Thus, receptors for light, smell, taste, temperature, sound, touch, position change, pressure, and tension are at hand which, as parts of the cybernetics system, can release physiologically meaningful protective reactions.

For example, temperature regulation works very precisely within certain limits. Hearing is also an extraordinarily sensitive sense organ: for the purpose of warning and information, it can receive acoustic vibration in the most important frequency range of 16–20,000 Hz. If, in addition, the ear could also detect the pressure fluctuations of the air as a result of molecular movement, this would be unnecessary and, because of the constant murmur, also troublesome. Unfortunately, the capacity of the ear to adapt to high noise levels – by changing the tension of the eardrum (tympanic membrane) – is too small to protect it from stress in the noisy technical environment. The physiological-acoustic regulatory system cannot provide sufficient resistance to these "unbiological" environmental conditions.

This statement applies even more to mechanical vibration, against which man can only apply an effective regulatory system under certain conditions (Dupuis 1969). Because of the lack of specialized receptors, there is no real sense organ for vibration (Keidel 1956). Instead, the perception of vibration must be regarded as the manifestation of a special excitation pattern in practically all of the mechanoreceptors in temporally changing periodic or nonperiodic excitation. The mechanoreceptors of the skin belong to this category (Meissner's and Merkel's corpuscles), which provide for skin surface reception. For vibration reception in more deeply situated tissue, the mechanoreceptors of the muscle spindles are responsible. The skin receptors can respond to both static pressure and changes in pressure of any given time function (i.e., vibration). In the muscle spindles, the receptors determine the position in addition to a time-dependent change in position.

Furthermore, the vestibular apparatus also serves for vibration reception, with the vestibule and the three semicircular canals for the position reception of the head and whole body (Schütz 1966). Here we have a division of labor:

through the maculae staticae, a vestibular system with crystalline corpuscles (statoconia), translational acceleration excites the vestibular nerves and allows information about position changes. Angular or rotational acceleration, i.e., of the head, is registered, however, by means of the flow of the endolymp relative to the canals through bending of the cupola (a crest with sensory cells).

Pain receptors can also signal the effect of mechanical vibration (Mandel 1963; Temple et al. 1964; Rublack 1978) when the vibration is of high intensity.

According to Keidel (1956), the transmission and central processing of the vibration information can occur on different levels. Thus, in the case of stored behavioral patterns (programming), reflexes can occur, i.e., the body responds with meaningful reactions. For example, this is the case in sinuslike or at least periodic vibration, under the influence of which the human body can better adjust and react (see section "Muscle Activity," p. 51). In particular, man can memorize, through experience, exactly defined vibration with a periodic course, as is exemplified in horseback riding.

In maintaining certain body positions, greater vibration stress can be tolerated. A skier who skis down a wavy, corrugated hill at great speed, for example, can maintain an almost stable position because of the vibration-reducing effect of his bent knees. This is made possible because the musculature chooses the effective spring and damping functions.

The body, however, cannot successfully process the vibration data received under all conditions. The reason for this is apparently that in the cybernetic system, the corresponding component parts are missing, for in the course of evolution there has been no necessity for a vibration-regulatory system that performs perfectly under all conditions. In particular, under the influence of stochastic vibration, and in a sitting position, these flaws in the sensory and motoric systems are unfavorably noticeable.

On the other hand, it should not be implied that in principle mechanical vibration has negative effects only on the human body. Undoubtedly, there are also positive effects, the influence of which is obviously based on the fact that vibration stress can cause the relaxation of certain tense conditions. However, in order not to achieve the opposite of what is being sought, the acceleration initiated must lie in a frequency range in which the human body is relatively insensitive. Its amplitude must not be too high, and the effective time must be kept short. Examples of such positive effects are vibration and massage apparatus which in a reactive manner may improve the blood circulation for a short period. One must nevertheless take into account the fact that when the exposure time is longer, a condition of monotony will be reached, which will lead to adaptation and result in reduced activity and fatigue. The baby cradle is an example, in that the rhythmic, slow movement promotes sleep through monotony. The rocking chair, on the other hand, as a work or resting chair, serves to change the body position continuously, particularly by changing the loading and unloading of certain parts of the skin and of the vertebral disks.

Occasionally, stimulation through mechanical vibration is also used for therapeutic purposes. For this reason, in 1969, Stankovic tried to stimulate intestinal motility by means of mechanical vibration in cases of intestinal atonia. He demonstrated, in 63 of the 100 patients examined, a pronounced increase in peristalsis after 30-s vibration exposure of 50 Hz to the right side of the lower abdomen. Cottet et al. (1968) have used mechanical vibration of 10 Hz at an acceleration amplitude of $3.5-9$ m/s^2 in order to induce expulsion of kidney stones.

There have also been many hypotheses formed and animal experiments conducted in order to investigate whether, with tension and pressure forces, mechanical vibration exerts a trophic stimulus to bone cells through which the cells grow, multiply, and in case of osteoplasts, strengthen bone structure (Roux's shaking theory, 1894). Experiments on rabbits (Sergl 1983), in which 100-Hz vibration was applied twice per day for 1 h to the lower jawbone over a 16-week period, showed local bone enlargement. However, the experiments are obviously insufficient to prove that the "vibration theory" is generally applicable. Sergl (1974, 1983) has compiled all the essential knowledge on this question, especially with regard to the publications by Whedon et al. (1949), Kummer (1959), Pauwels (1960), Häupl (1955), Jankovich (1972), Oates et al. (1978), and Shapiro et al. (1979). In this survey many contradictions are apparent and no one theory can explain all of the results in detail. Therefore, so far no general proof has been provided that vibration can enhance bone growth in the sense of a therapeutic measure.

In addition to the aforementioned examples of positive effects, today unfortunately the disadvantageous effects caused by mechanical vibration must be taken into account. From the knowledge available on the reaction of the human body to mechanical vibration, one can assume that man was not "constructed" by nature to suffer the kind of vibration that assaults him today in this technological era. Above all, this applies to the condition when man is exposed to mechanical vibration in a sitting position, as in modern vehicles, or through the hand-arm system.

The prerequisite for any preventive measure is first to acquire knowledge on the biodynamic reaction to vibration of the body and its parts, as well as of the physiological effects and the health-damaging long-term effects of such vibration stress.

5.2 Biodynamic Reaction on Vibration

5.2.1 Models of the Human Body under Vibration

Knowledge of the reaction of the body and its parts to mechanical vibration is useful in order to deduce the extent of the physiological and pathological reactions to be expected, as well as the strength of perception; conversely, it is also useful to be able to interpret such reactions from a biomechanical point of view. It is especially important to recognize the resonance phenomena because − like technical material and structure − the biological tissue and parts of the body are placed under especially strong stress when they are in resonance.

In the literature there have been many attempts to consider man and his reaction to vibration from the mechanical point of view in order to form theoretical models for the human body as a vibration system, and to use them for mathematical calculation of reactions (Haack 1953; Dieckmann 1957, 1958 a; Goldman and von Gierke 1960; Coermann et al. 1960; Coermann 1961, 1965; Sassor and Krause 1966; von Gierke 1968; Serbitzer 1974; Sandover 1978; Vogt et al. 1978; Mertens 1978).

The prerequisite is knowledge of the reaction of man and different parts of his body to vibration. As an object capable of vibration, the human body consists of many different kinds of tissue; their strength and elasticity properties differ widely from each other. Composed of more than 200 single bones, the skeleton serves, among other things, to support its own body weight and to transmit forces. Different types of joints are provided for the various articulations and the mobility of the skeleton; the thin cartilage layers of the joint surfaces are covered by the synovial fluid. Elastically stretched tendons serve to bind the joints together. The spring elasticity of the joint junctions is dependent on the kind of joint tissue (soft cartilage, hard cartilage, connective tissue). In the spinal column, the disks are the elastic joint connections, which also have certain elastic and damping properties.

The musculature that stretches over the joints makes movement of the joint parts possible, as well as the transmission of forces. Depending on the magnitude of the forces, the muscle tissue must contract, either statically or dynamically, with variable strength. Therefore, the musculature can counteract and resist vibratory forces.

The internal organs and soft tissue (except for muscles) and body fluids are, however, incapable of actively reacting to vibration strain, but remain passive, i.e., they are subjected to vibration-caused mechanical strain.

The exterior soft tissue can attenuate vibration of small amplitudes and high frequencies transmitted to the skeletal system; the internal organs are elastically suspended in the volume provided by the abdominal wall; the diaphragm represents independent, separate vibration systems.

To represent the human body by a theoretical model, the main body segments are modeled as spring-mass systems with dampers and are connected to a complex system. This overall complex elastic system can be very complicated, however, because of the many different effective masses involved (body tissue of different kinds) and the various elastic and damping properties (muscles and tendons). Although the masses are approximately known from anatomical data, the spring and damping values change considerably, especially as muscle tension varies according to body posture.

This complicates the physical-mathematical approach to models of this kind and makes experimental data on biomechanical vibration reaction mandatory.

Special difficulties arise when it is desirable to know the vibration reaction of a certain part of the body, a "subsystem"; by the measuring techniques available, this information is either unobtainable or difficult to obtain. For this reason, frequently only the dynamic properties of the main body parts are represented in such models.

Mechanical models for the seated person have been proposed involving one or more spring-mass-damper systems (Fig. 2). Dieckmann (1958a) thinks that, for many practical purposes, a one-spring one-mass substitute model is sufficient for the frequency range up to approximately 10 Hz, as this includes the main resonance of the human body. He has also established a second resonance range of the human body between 10 and 13 Hz, which can be modeled with good approximation with the help of a two-mass two-spring system. Sassor and Krause (1966) point out, however, that even the two-mass two-spring system is only an approximation. The seven-mass seven-spring system could certainly come very close to reality. Calculation of this model would be very complicated, however, particularly if changing muscular tension is taken into consideration. Coermann (1965) asks that we remember that the subsystems are no longer accessible singly and that their parameters cannot be directly measured. He recommends that, for a certain kind of strain, the dynamic properties be determined experimentally for the whole body.

Because most of the parameters that have an influence are unknown, the strain in subsystems of the body cannot be calculated. Therefore, experimental tests have been carried out by many authors to establish the vibration reaction of certain parts of the body, as well as of the body as a whole. For this purpose, force, displacement, velocity, and acceleration have all been measured.

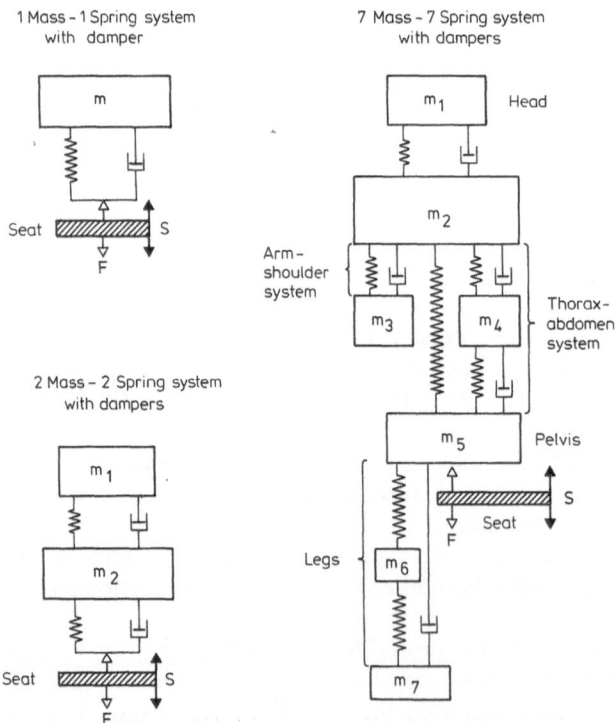

Fig. 2. Mechanical vibration models for the human body in the seated posture [adapted from Dieckmann (1957, 1958) and Coermann (1963, 1965)]

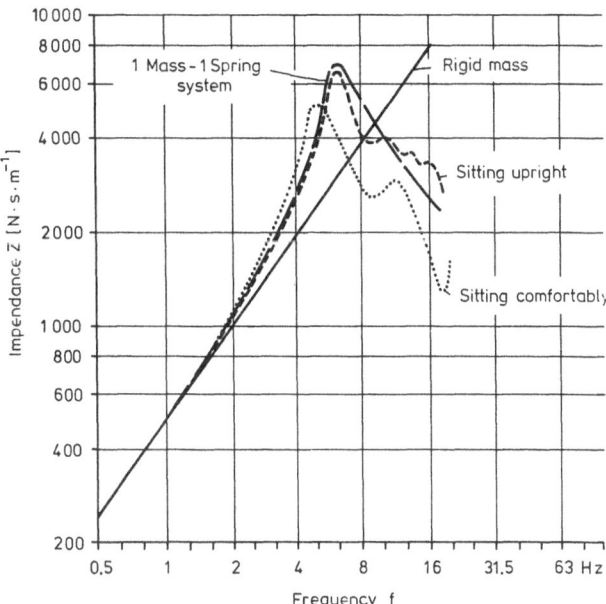

Fig. 3. Impedance Z of a subject (840 N) in various postures. Comparison with a rigid mass and a one-mass one-spring system (z direction). From Coermann (1963)

Several authors have experimentally determined the mechanical impedance, Z, of the human body (as the input or driving point impedance in contrast to transfer impedance in different parts of the body): this impedance is given by the quotient of force F transmitted from the seat to the human body, divided by the vibration velocity, v, at the site of force transmission:

$$Z = \frac{F}{v} \ (N \cdot s \cdot m^{-1}) \,.$$

From measurements by Coermann (1963), Fig. 3 shows that only in the frequency range below 1.5 Hz can the transfer of force on the human body basically be determined by mass and that acceleration is proportional. Over 1.5 Hz, with a maximum at 6 Hz, vibration exposure leads to resonance of the human body when seated in an upright position (relaxed already at 5 Hz). Up to this maximum, a good correspondence has been found with the one-mass one-spring system. But above 5 or 6 Hz, the curves deviate; this is caused by the secondary partial resonance of the torso, which occurs, depending on muscle tension, in the frequency range 10–15 Hz (also see Vogt et al. 1968).

The application of impedance curves for the human body is mostly useful for the calculation of theoretical models of vibration-protection systems like suspended platforms, seats, and vertical suspension. In any case, this is the main application mentioned in the international standard ISO 5982 (1981), for frequency-dependent impedance curves for the standing, seated, and reclining human body (see Figs. 4–6). It is pointed out that standardized impedance curves

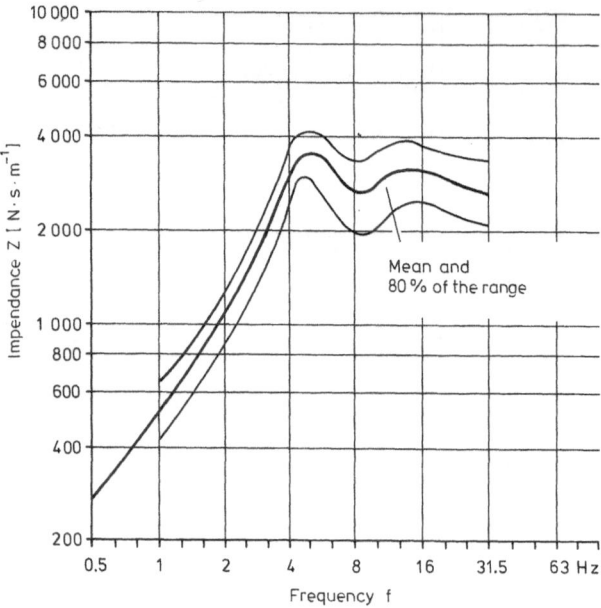

Fig. 4. Impedance of the human body in the standing posture (ISO 5982)

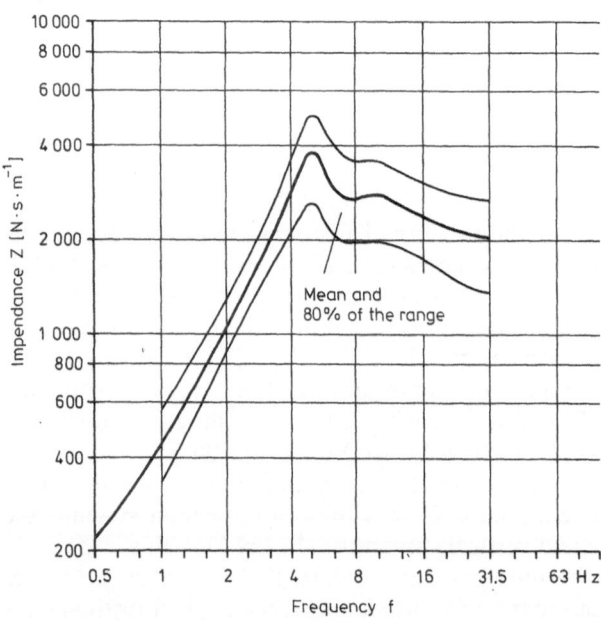

Fig. 5. Impedance of the human body in the sitting posture (ISO 5982)

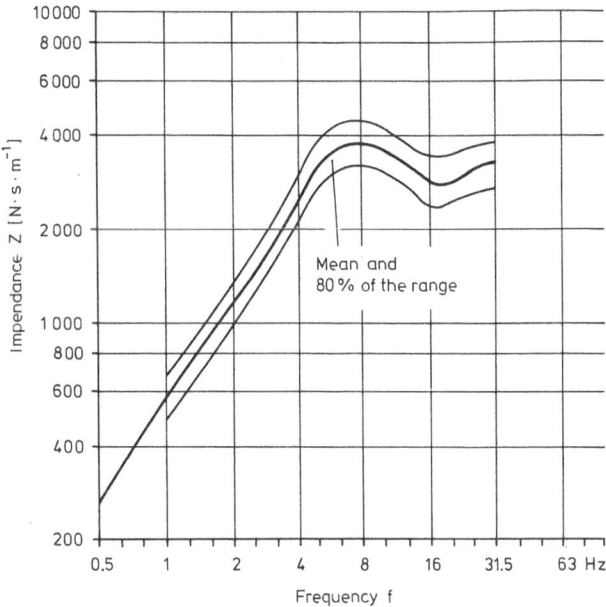

Fig. 6. Impedance of the human body in the reclining position (ISO 5982)

are only applicable to linear systems and do not reflect the properties of subsystems.

To judge the vibration reaction of single parts of the body (e.g., the spinal column, internal organs), whole-body impedance measurements are therefore not suitable. Furthermore, there is another restriction: with regard to impedance measures, the system being tested, the human body, must remain in constant contact with the vibration-created transmitting surface; thus, the body must not be removed from the seat or floor. This, however, cannot be guaranteed when there is extreme stress, such as on tractors, building machines, and military vehicles.

Another characteristic of impedance measurements is that the parts of the body that are close to the place of entrance of vibration (e.g., pelvis, lumbar vertebral column) influence the impedance curve to a greater degree than the body parts that are further away (e.g., cervical vertebral column, head).

Although in the following chapters more detailed descriptions will be provided on the reactions of the human body to vibration while walking, standing, sitting, and reclining, in this passage we will confine ourselves to presenting impedance curves that are standardized according to ISO 5982 (Figs. 4–6). The ranges indicated include approximately 80% each of the underlying experimental data, the scattering of which is partly caused by the differences in body weight. Variable body postures also play a role, and these were frequently not precisely defined. All data have been measured by applying sinusoidal vibration stress with acceleration amplitudes of $1-2.5$ m/s² (Figs. 4 and 6), or $1.0-2.0$ m/s² (Fig. 5).

5.2.2 Vibration Behavior when Walking and Running

Walking and running are natural means of locomotion. In particular, one can assume that the vibration stress that occurs, in connection with the work performed by the muscles, normally does not have damaging consequences. In order to obtain an indication of the vibration characteristics of the human body in an upright position, it is interesting to compare man's reaction to vibration with regard to gait. For this purpose, comprehensive experiments on locomotion have been carried out in the fields of biodynamics, orthopedics, and sports medicine. However, the goals of these experiments were often different from the ones that interest us here.

Simic (1970), Rao and Jones (1975), Dupuis et al. (1975), Barton (1981), and Dielmann (1983), among others, have studied problems dealing with the vibration-transmission behavior of man when walking. As far as the various measurement and evaluation procedures permit comparison at all, the measurement results are all in approximate agreement. However, there are differences in the interpretation of the results, particularly when the interpretation is made solely from a dynamic point of view with evaluation of the physiological background.

The goal of an experiment by Dupuis et al. (1975) was to acquire data on vibration stress in the immediate vicinity of the eyeball during various kinds of locomotion on foot or by other means. An attempt was made to deduce by comparison and estimate the risk of detachment of the amotio retinae (retinal detachment). For this purpose, acceleration was measured at the place of entrance of vibration in the body and at the forehead during 31 different means of locomotion. Table 4 shows samples of the results found. It has been found that the acceleration occurring in the head during walking is of the same order of magnitude as when riding in vehicles. Great vibration stress takes place especially when walking rapidly upstairs, running track, riding on horseback, or riding in a motorboat over high waves. Nigg and Neukomm (1973) determined $5-20$ m/s^2 as the median acceleration on the head when skiing and $10-20$ m/s^2 when running in tennis shoes in a gymnasium or barefoot on asphalt, or $10-30$ m/s^2 with spikes on an artificially surfaced track. As a result, it was concluded that the ophthalmologist should advise patients at risk of retinal detachment to take the following precautions (Draeger and Dupuis 1975):

1. To protect the eye from overstrain from mechanical vibration, the following situations should be avoided: rapid walking, running, sprinting, use of vehicles with poor suspension on uneven roads, downhill skiing, horseback riding, waterskiing, and riding in motorboats.
2. Since vibration stress sharply rises with increased speed, each type of locomotion on foot should be done at a moderate speed.
3. For footwear, generally shoes with particularly soft soles are recommended.
4. The elastic quality of the walking surface, e.g., soft carpets, also probably reduces vibration stress.
5. High body weight increases vibration stress: overweight should be avoided.

An additional experiment carried out by Dielmann (1983) was designed to clarify vibration transmission to certain sections of the spinal column, while

Table 4. Vertical acceleration at the head (orbita) and transmission factors in various types of locomotion (Dupuis et al. 1975)

Type of locomotion	Vertical acceleration at the head	
	rms acceleration (m/s^2)	Transmission factor V[a]
A. Walking/running		
Walking on grass	1.9	0.36
Walking on asphalt streets	2.3	0.47
Walking on cobblestones	3.0	0.47
Walking rapidly downstairs	4.0	0.40
Running on a cinder track	12.5	0.32
B. Sitting posture		
Bicycle riding on asphalt streets	1.6	–
Bus riding on good roads	1.5	1.25
Small cars on poor roads	2.1	1.62
Tractor driving on field roads	2.8	1.14
Motorboats in waves	8.6	1.21
Horseback riding at a gallop	9.5	1.28

[a] Transmission factor $V = \dfrac{\text{Acceleration at the head}}{\text{Acceleration at the entry site in the human body}}$

simultaneously recording the muscle activity during several kinds of locomotion. With regard to transmission factors from the place of entrance of vibration in the body to the places measured on the body, such as the lumbar or cervical parts of the spine, the results were basically the same as presented in Table 4 (head). While engaging in natural kinds of locomotion like walking or running, the lower extremities (legs) proved to be an extraordinarily effective spring-suspension system: vibration was reduced by 30% – 70%. In a sitting position, the vertebral column has less chance of reducing vibration. When using most vehicles or when horseback riding, the result is instead increased vibration with a transmission factor above 1.0. This has to do with resonance vibration of the trunk as a whole and of sections of the vertebral column (also see section called "Vibration Behavior in Sitting Posture" and "Vibration Behavior of the Spinal Column" on pp. 26 and 31).

In the kinds of locomotion under examination (walking and running), the absolute values of vertical acceleration in the areas of the lumbar and cervical vertebrae are approximately the same as the ones that occur when operating a wheel loader, for example. Nevertheless, the strain to the vertebral column from both kinds of stress cannot be equated, because the muscular reaction also has to be taken into consideration. When sitting in vehicles and working machines, the vibration stress, because of the resonances mentioned, continuously compresses and stretches the intervertebral disks. Junghanns (1979) suspects that through this mechanism, the diffusion may be disturbed, with a resulting reduction in the nutrition of the invertebral disks. As Dielmann (1983) showed by means of electromyography, no effective dynamic muscular activity of the *m. erector spinae* occurs during natural movement in the upright position (walking and running especially). He interpreted this with regard to vibration strain as "protective activity." Vibration stimulation in natural kinds of locomotion is different from that in vehicles. This stimulation and the muscular fixation of the vertebral system (with the trunk as a mass-spring system) hinder the excitation of resonance in the area of the vertebral column.

5.2.3 Vibration Behavior in Standing Posture

5.2.3.1 Vertical (z-Axis) Excitation

With the help of impedance measurements, Dieckmann (1958a) has established the frequency-dependent vibration reaction of the standing person exposed to vertical z-axis vibration. The results are included in the international norm ISO 5982 (Fig. 4). The first resonance position is established at approximately 5 Hz; the second is smaller at about 12 Hz. Below the resonance frequency where the impedance has primarily the characteristics of a mass, Dieckmann has not been able to establish any real differences between standing and sitting, as the mass of a person sitting or standing is the same. In the resonance region, the height of the maximum and its exact position are dependent on the body posture when standing (upright, slightly or deeply bent in the knees). Thus, a downhill skier who skies, for example, at high speed downhill over uneven terrain will respond with his knees to the unevenness of the surface. The elastic response can be improved by deeply bent knees, and resonance vibration can be avoided. Miwa (1975) has investigated six variations of the standing posture.

Goldman and von Gierke (1960) were able to confirm both of the above resonance ranges on the basis of Coermann's experiments. Miwa (1975) found a clear average resonance of 7 Hz, using 20 experimental subjects in the upright standing position. This resonance did not vary appreciably with differing body weight.

5.2.3.2 Horizontal (x-Axis, y-Axis) Excitation

In the horizontal dorsoventral vibration direction, x, Dieckmann (1958b) measured acceleration at the knee, hip, shoulder, and head in the standing subject. He showed clearly that there is no resonance vibration in the body in this direction, in contrast to the vertical vibration direction (Fig. 7). In this figure the displacement amplitudes calculated from the acceleration measurements are represented by the main axis of the ellipses. The time shift between the vibration (phase) is expressed by marking corresponding points on the circumference of the ellipses and by connecting them with straight lines. The ellipses do not reflect the vibration form in this figure, but are used to illustrate the phase shift between body parts.

Fig. 7. Amount and phase of the vibration amplitudes at various parts of the body in the standing posture during horizontal vibration in the x direction (Dieckmann 1958 b)

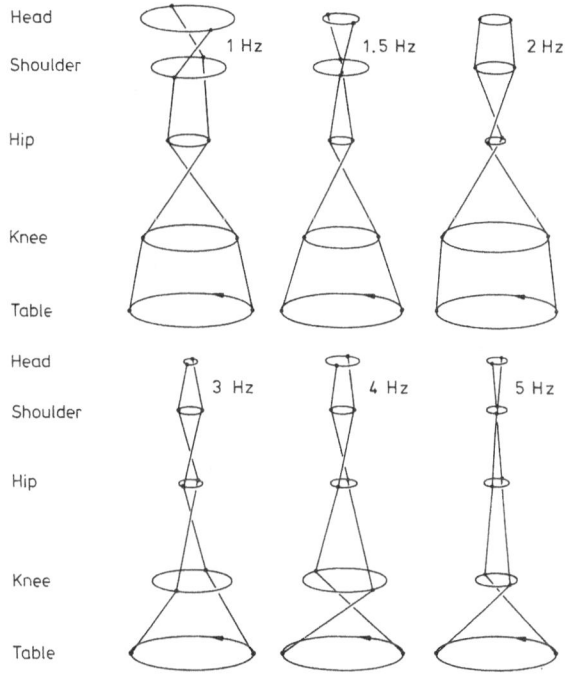

In Fig. 7 it is apparent that, as the frequency increases, the phase shift between all parts of the body and the vibration platform also gets larger. This is apparent in the shift of the location of the nodes.

In the opinion of Dieckmann (1958 a), during horizontal stimulation in the standing posture these nodes permit the processes to be regarded as "standing waves." The experiments have proven that between 2 and 3 Hz a phase shift of approximately 360° occurs at the head, so that the head vibrates again in phase with the vibrating platform. There is no vibration increase compared to the platform, however, since the head in the standing position is far away from the excitation site. The relatively lowest vibration amplitudes occur in the area of the hips. On the basis of the natural frequencies of different body parts, Dieckmann (1958 b) considers the frequency range between 1 and 3 Hz − corresponding to 4 − 5 Hz in vertical stimulation − as particularly severe. He carried out experiments in the y-direction with results not appreciably different from those in the x-direction.

5.2.4 Vibration Behavior in the Reclining Posture

The vibration reaction of the reclining human body is of interest when passengers are exposed to mechanical vibration while being transported in sleeping or couchette cars of railroads, on ships, in the cabin of a truck or, if sick or injured, in an ambulance, a helicopter, or an airplane. Half-reclining body positions are of interest in luxury travel buses or in space travel.

For these reasons, vibration-reaction experiments on reclining or half-reclining body positions have been carried out by several authors. The particular
feature shared by all these experiments is that the body is not merely exposed
to vibration in a limited area, as takes place in a sitting or standing position, but
that the vibration generally excites large areas of the body, the back of the
head, the trunk, and the legs. The vibration-transmission behavior must, therefore, be determined simultaneously and evaluated in different parts of the
body.

5.2.4.1 Vertical (x-Axis) Excitation

Among others, Vogt (1973), Vogt and Krause (1973), and Vogt et al. (1978)
have carried out impedance measurements on the reclining person under vertical vibration stress x. The results of the experiments performed by Edwards and
Lange (1964), Fairbands (1964), Coermann (1964), Vogt and Krause (1973),
and Vogt et al. (1973) all indicate that the main resonance of the whole body is
in the range of approximately 5−8 Hz. This fact has resulted in impedance
curves that have been standardized according to ISO 5982 (Fig. 6). Weis and
Primiano (1966) have determined, however, that because of resonances of parts
of the body, resonances of the reclining body occur up to approximately 80 Hz.
   Szameitat (1976) and Szameitat and Dupuis (1976) have examined vibration-transmission behavior in the reclining person whose ankles, knees,
stomach, sternum, and forehead were stimulated by vertical vibration. Their results are based on statistics from experiments on 14 test subjects. The summarized results are shown in Fig. 8. It is obvious that the vibration behavior is
similar for the knee, stomach and sternum, that that of the feet is lower, and

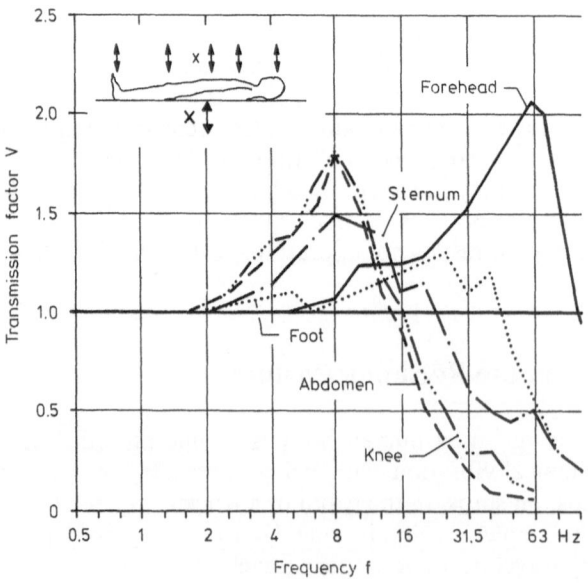

Fig. 8. Frequency-de-
pendent vibration trans-
mission at five parts of the
body in the reclining posture during vertical vibration in the x direction
(Szameitat 1976; Szameitat
and Dupuis 1976)

that that of the forehead has a particularly great maximum at approximately 63 Hz. Consequently, a high sensitivity in the frequency areas of about 4–10 Hz and 30–80 Hz can be expected.

5.2.4.2 Horizontal (y-Axis, z-Axis) Excitation

Because of the considerable mobility of the supine body on a smooth surface during vibration stimulation in the horizontal directions, y and z, resonances occur at low frequencies with greater amplification than for the vertical direction. This finding has been established by Coermann et al. (1960) by means of acceleration measurements of forced chest expansion and respiratory volume for the z-direction. Szameitat (1976) and Szameitat and Dupuis (1976) have been able to determine the transfer functions to the forehead, abdomen, and foot for both horizontal vibration directions, y and z (Figs. 9 and 10). These frequency-dependent functions show a similar course, even though there was a shift to slightly higher resonance frequencies in the direction of the vertebral

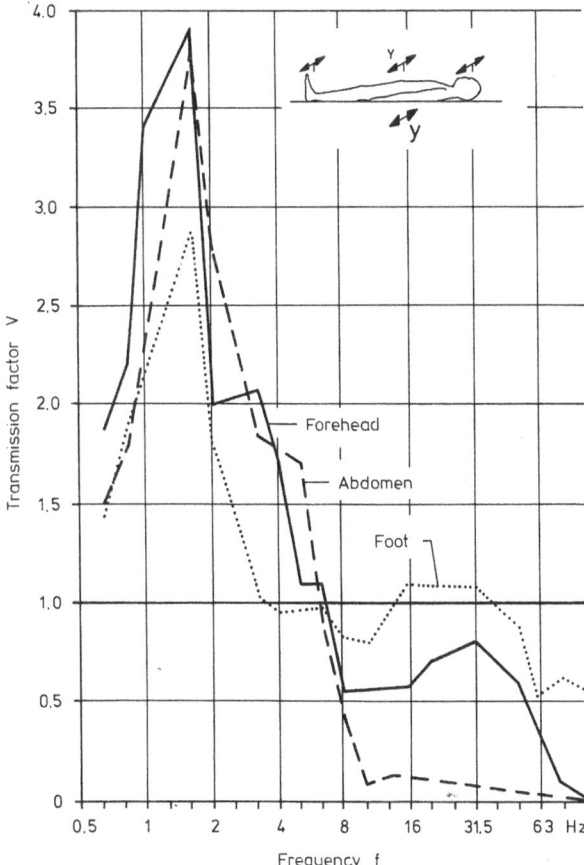

Fig. 9. Frequency-dependent vibration transmission at three parts of the body in the reclining position under horizontal vibration y (Szameitat 1976; Szameitat and Dupuis 1976)

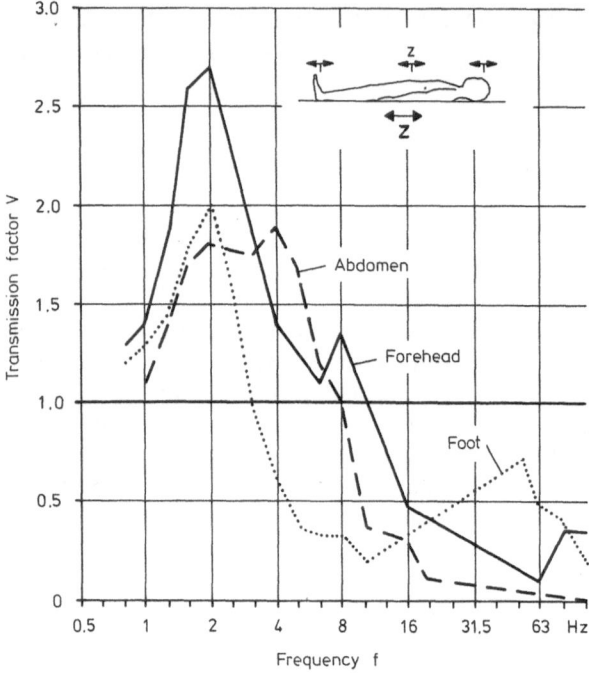

Fig. 10. Frequency-dependent vibration transmission at three parts of the body in the reclining position under horizontal vibration z (Szameitat 1976; Szameitat and Dupuis 1976)

column, z. In the y-direction the resonance maximum is about 1.7 Hz, with transmission factors of 3.0−4.0 (Fig. 9); in the z-direction it is 2−4 Hz, with transmission factors of 1.8−2.7 (Fig. 10), which decrease greatly with further increasing frequencies.

In other experiments in which the subjects were imbedded in stiff body molds ("evacuated mattresses"), as used in ambulances, highly effective stabilization could be shown in the horizontal vibration direction. This is of great significance for transporting the sick and injured (Dupuis and Hartung 1981).

5.2.5 Vibration Behavior in Sitting Posture

Most experiments have examined the biodynamic vibration behavior of the human body while seated. There are probably two reasons for this. First, this position is encountered most frequently compared to other positions. Second, this type of exposure is also responsible for most vibration-caused complaints, disturbances, and injuries to health. Current knowledge in this area will therefore be presented in greater depth − particularly current knowledge regarding the behavior of individual parts of the body.

5.2.5.1 Vertical (z-Axis) Excitation

Among others, Dieckmann (1956), Lehmann and Dieckmann (1956), Coermann (1961), Coermann and Okada (1964), Vogt (1968), Miwa (1975), and Mertens (1978) have all measured impedance under vertical z-axis vibration stress in the seated position. Their results are included in international standard ISO 5982 (1981; Fig. 5).

Even more experiments have determined frequency-dependent transmissibility by means of acceleration measurements at the seat and at exterior sites of the body. Included in these experiments are those by Müller (1939), Dieckmann (1956, 1957), Schmitz (1959), Guinard (1960), Coermann (1965), Berthoz (1966), Winkelholz (1967), Dupuis (1969), Shoenberger (1972), Griffin (1975a), Zagorski et al. (1976), Rowlands (1977), Griffin and Whitham (1978), Griffin et al. (1978), and Vogt et al. (1979). Specific experiments on the vibration behavior of the head have been carried out by Berthoz et al. (1972) and Vogt et al. (1979). All these results will go into a new edition of the ISO standard 2631 still in preparation.

Figure 11 shows the summarized results from experiments on vibration transmission from the seat to the head, which have been carried out with various exciting acceleration amplitudes. It is apparent that above 1.5 Hz, transmissibility factor V increases steeply to $4-5$ Hz, only to decrease then equally steeply, so that above 7 Hz there is a marked decrease in vibration. The vibration amplification in the resonance area of 5 Hz is $V = 1.8-2.0$.

Transmission functions have also been determined in the shoulder-neck area during vibration stress in the seated position (Lehmann and Dieckmann

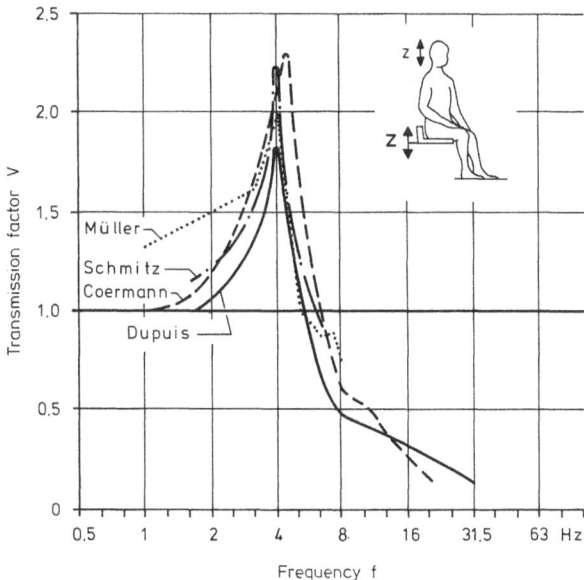

Fig. 11. Frequency-dependent vibration transmission from the seat to the head in the vertical direction z (Müller 1939; Schmitz 1959; Coermann 1963; Dupuis 1969)

1956; Dieckmann 1956; Coermann and Okada 1964; Dupuis 1969; Zagorski et al. 1976; Rowlands 1977). Whether the vibration is greater in the shoulder-neck area or on the head is not clearly determined by these studies. Apparently, the results depend on the exact measurement site and the type of body posture. Thus, according to Dieckmann (1956), measurements obtained directly at the shoulder result in greater values than those at the head because of the considerable mobility of the shoulder in the resonance area, i.e., between 2 and 5 Hz. However, if one looks at measurements obtained close to the cervical part of the vertebral column − over C3, for example − one finds that the resonance-conditioned increase in vibration is not as great at this site as at the head (Dupuis 1969). Above the resonance area of > 6 Hz, however, vibration at the head is appreciably lower than at the cervical vertebral column (Dupuis 1969; Zagorski 1976). This finding indicates that above the natural frequency of the trunk, the vertebral column increasingly acts to absorb the vibration. The spinal column, aided by the elasticity of the intervertebral disks of the cervical vertebral column, alternates vibration in the caudocranial direction.

In this connection, it should be pointed out, however, that head measurements often cannot be compared if different measuring sites have been chosen. The head movements are always "complex," i.e., they consist of rotation and linear movements.

Furthermore, the influence the kind of body posture in the seated position has on vibration transmissibility at the head is also of practical significance. Corresponding experiments have been made by teams such as Coermann and Okada (1964), Griffin (1975a), and Dupuis and Hartung (1980a). Griffin had his test subjects assume two different body postures, one of which produced the lowest vibration sensation in the head area, and the other the highest, relatively speaking. He found that vibration stimulation at 7 Hz resulted, for the two body postures, in differences in acceleration at the head of approximately 100%, which increased further with increasing frequency.

Dupuis and Hartung (1980a) examined the "natural" posture (relatively loose, relaxed, and somewhat limp) and the straight, erect body posture, while the subject was seated without leaning against a backrest. Three different acceleration intensities were used at logarithmically spaced frequencies (Figs. 12 and 13). At the resonance frequency of 4 Hz, there was on average a stronger increase in amplitude in the erect seated position than in the natural, relaxed seated position. However, in the resonance area, the differences are difficult to determine statistically. Above the resonance area of > 8 Hz, vibration transmissibility to the head in the erect body position is considerably greater and differs significantly from the natural posture.

With regard to the influence of acceleration intensity, the highest vertical acceleration (4 m/s²) resulted in the greatest amplification at resonance.

In the experiments by Coermann and Okada (1964), the seat-back angle, with respect to the horizontal seat surface, was varied (8 steps from 90° to 140°). When the incline of the backrest was 90° − 105°, they established that the vertical vibration transmission to the head when leaning against the headrest was practically constant, but that it decreased when the incline was increased to 140°. At the same time, they pointed out that although the passenger

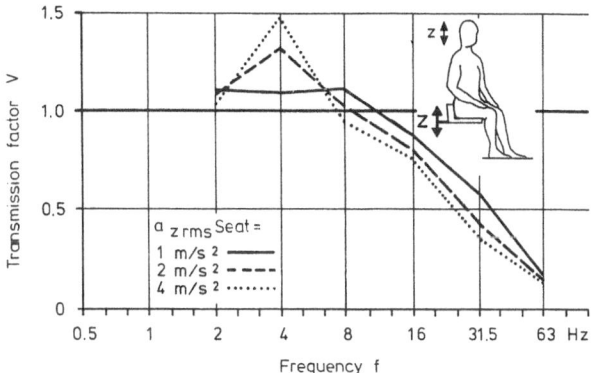

Fig. 12. Frequency-dependent vibration transmission from the seat to the head in the upright posture during vibration in the vertical direction z (Dupuis and Hartung 1980a)

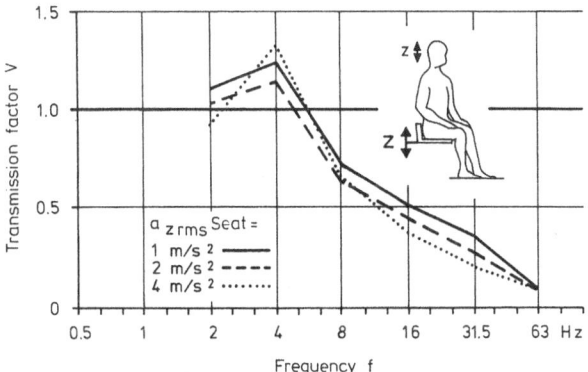

Fig. 13. Frequency-dependent vibration transmission from the seat to the head in the "natural" posture during vibration in the vertical direction z (Dupuis and Hartung 1980a)

benefits from a steeply inclined backrest, the driver may not use this position for ergonomic reasons.

In general, the type of body posture in the sitting position can be considered to be of great influence regarding vibration transmission. However, at work the body posture is often determined by the specific kind of task to be performed.

5.2.5.2 Horizontal (x-Axis, y-Axis) Excitation

Far fewer experiments have been conducted on the effects of horizontal mechanical vibration in the sitting position − probably because these effects seem to have much less influence than those in the vertical direction. Dieckmann (1958b) established that when there is vibration stimulation in the horizontal x-direction, vertical vibration components occur simultaneously, resulting in ellipsoidal movements of the head. Dieckmann believes that these movements are connected with the problem area of kinetosis.

Dupuis and Hartung (1980a) have examined the reaction of the head when excited with horizontal vibration. Six logarithmically spaced steps were used (2−63 Hz) and three intensity steps ($a_{eff} = 1$, 2, and 4 m/s^2) in the x- and y-directions. Since there is not much difference in the results of the two directions, only the y direction results are described (Figs. 14 and 15). The horizontal y- and vertical z-vibration components of the head were included in the examination. The results of Figs. 14a and 15a show that there is no amplification of the horizontal vibration at the head. The vibration components occurring simultaneously in the z-direction (Figs. 14b and 15b) are small − approximately half the size of the vibration in the y direction. Above 8 Hz the acceleration at the head is considerably reduced.

No significant differences could be determined in the vibration behavior between the upright (erect) and natural (relaxed) body posture. These data are valid for the vibration directions, y and x, as well as for the vertical vibration component, z, of the head.

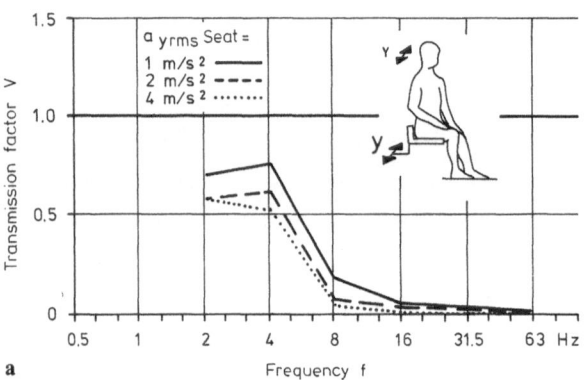

a

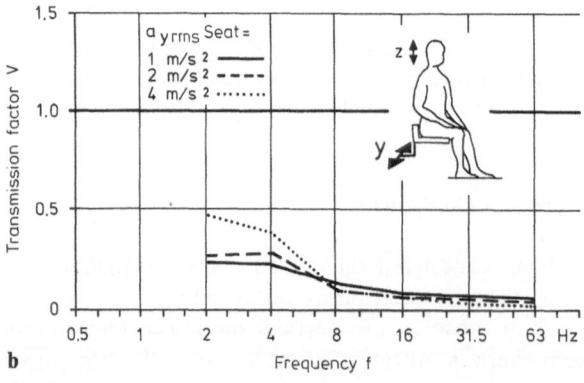

b

Fig. 14a, b. Frequency-dependent vibration transmission from the seat to the head in the "upright" posture during vibration in the horizontal direction y (Dupuis and Hartung 1980a). **a** y components; **b** z components

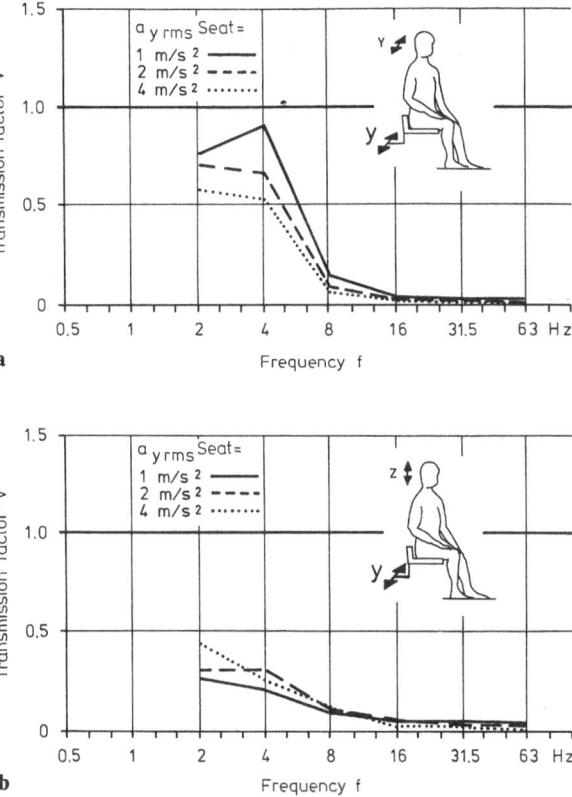

Fig. 15 a, b. Frequency-dependent vibration transmission from the seat to the head in the "natural" posture during vibration in the horizontal direction y (Dupuis and Hartung 1980 a): **a** y components; **b** z components

5.2.6 Vibration Behavior of the Spinal Column

The biodynamic vibration behavior of the trunk while walking, standing, and sitting, has already been dealt with in previous sections, in which measurement results were also established in the immediate vicinity of the vertebral column. It is also interesting to know how individual disks or segments of the spine react alone or in relationship to one another when under the influence of vibration. Since these considerations allow conclusions to be drawn about the strain involved (see section "Chronic Effects of Whole-Body Vibration," p. 87ff), this question will be dealth with in detail.

In the German-speaking countries, the most comprehensive description of the influence of occupation on the vertebral column and the biodynamic and biochemical problems of stress on the vertebral column has been published by Junghanns (1979). This description also includes the effects of mechanical vibration, as well as numerous citations from the literature. Farfan (1979) has investigated in particular the biodynamics of the lumbar vertebrae, as this area is especially affected.

After visiting the most important research institutes working in this field in Europe and the United States, Sandover (1981) compiled the newest findings on strain of the vertebral column in truck drivers. His research report contains voluminous citations from the literature and presents unclarified scientific questions for further study.

In view of the apparent strong danger to the vertebral column during intensive vibration stress, the vibration behavior of the vertebral column will be studied more closely later.

Static stress experiments on dissected intervertebral disks and vertebral bodies have been carried out by Brown et al. (1957) and Hirsch (1965). Discometric experiments were undertaken by Nachemson and Morris (1964), using 16 subjects who were exposed to an important experimental procedure. They were able to show that pressure on the invertebral disk L 3−4 in the sitting position is almost 40% greater than when in the standing position. Horst (1982) made a further important contribution to the understanding of the mechanics of the intervertebral disks: for the first time, it was possible to measure the forces on the end plates of the vertebral bodies. From these results, the relatively great preloading of the vertebral column of drivers sitting in road vehicles and work equipment can already be recognized in the static condition.

Only a few experiments are known in which the dynamic behavior of the vertebral column has been investigated when under the influence of vertical vibration. Krause (1963) has studied the effect of vertical vibration in a frequency range of 10−40 Hz at a constant stimulus amplitude of only 0.8 mm; two different procedures were applied to only one subject. The relative vibration of neighboring vertebral bodies in the vertical direction was determined by applying small tubes to the spinal processes on a lean subject from the 2nd to the 5th lumbar vertebrae and connecting them with a circular strain gauge. However, the bending and torsion vibration of the sacrum to the 4th thoracic vertebra was measured optically with the help of a galvanometer mirror.

In the area of the lumbar part of the spine, the bending and torsion angles were appreciably greater than in the dorsal region. The greatest vertebral disk deformation and the greatest vertical displacement were established between the L 3−4 vertebrae. During bending, the vertebral disks were most strongly deformed at 10 Hz and 35−40 Hz and during torsion, at 20 Hz and 32−40 Hz. It was found that the lumbar region of the spine is under especially high stress, whereas the thoracic part is stiffer and therefore exposed to less bending and torsion. This finding corresponds to the results of Bakke (1931) who investigated the movement of individual vertebral bodies against each other.

Lange and Coermann (1965) refined this measuring procedure by using a mercury device to measure the change in distance between two spinal processes. They investigated the vibration behavior of vertebral segments L 4−5 and L 3−4 in a test subject. At a constant acceleration, beginning with 5.0 m/s², there were relative shifts of neighboring vertebral bodies of around 0.05 mm (at 12 Hz) to 0.6 mm (at 4−5 Hz).

In addition to the experiments carried out, it appeared relevant to study the effect of vertical vibration at lower frequencies (< 8 Hz), which are the most important with regard to the behavior of the spinal column in vehicles. The results of the aforementioned studies, as well as experience in orthopedic practice, indicated that, above all, the parts of the lumbar region, in addition to the cervical region, should be included in such experiments (Christ and Dupuis 1966; Dupuis 1966; Dupuis and Christ 1967). Although the greatest demands are placed on the cervical area of the spine with regard to mobility (Bakke 1931; Buetti-Bäumel 1954), the greatest dynamic forces must be transmitted through the lumbar area due to the masses involved.

When conducting preliminary research, Christ and Dupuis (1963) found, for observation of the cervical vertebral area, roentgenocinematography methodologically suitable when using a 12.5-in. screen and an electro-optic image intensifier. In addition, the equipment permitted the person conducting the experiment to view the running X-ray pictures in a television monitor without darkness adaptation and thus be able to control the photographs of the experiment, as well as record film shots with 48 pictures/s on 35-mm film (Fig. 16). Radiation-protection requirements were strictly adhered to with regard to personal protection and the dosage to which the test subjects were exposed. Because of the film-speed limitation of 48 pictures/s, vibration frequencies of only up to 6 Hz could be investigated.

In order to study the behavior of the cervical spine during vertical sinusoidal vibration, with frequencies between 0.5 and 5.5 Hz, a series of research projects were carried out with seven subjects. Each series had a constant displacement amplitude (peak value) of 7, 10, and 20 mm.

Representation of the lumbar region of the spine with roentgenocinematography presented difficulties because the shots were partially masked from the front (ventrodorsal) and side and because of the increased radiation dosage required. For this reason a method was developed that permitted the movement

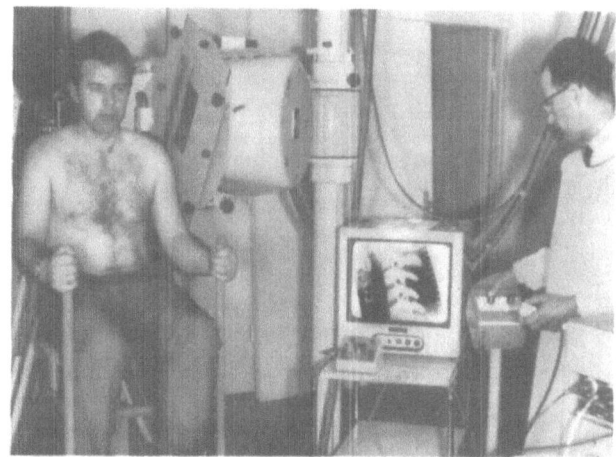

Fig. 16. Viewing the cervical spine under the influence of vibration by roentgenocinematography through a television monitor

of the lumbar area to be observed with optical film shots. Because of lack of practical and technical experience with the complicated procedures of Krause (1963) and of Lange and Coermann (1965), it was decided that these procedures would not be used. Another procedure was used instead. The Assistant Medical Director, Dr. W. Christ, an orthopedist, used himself as experimental subject and marked the spinal processes of the dorsal vertebra T12 and the lumbar vertebrae L 1 – 4 by implanting wire pins.

The skin and connective tissue over the spinal processes were anesthetized locally and wire pins were bored about 15 mm into the spinal processes of the five vertebrae and allowed to protrude about 100 mm. The wire pins had to be parallelly aligned. The wires bored in at T12, L2, and L4 were also marked at their tips with black and white paint in order to permit good visualization of movement in the sagittal plane in film shots (Fig. 17). With the subject prepared in this manner, various experiments were carried out over a 4-h period.

During observation of the pin-marked lumbar vertebral area under vibration stress, with a vibration input displacement of 5 mm as peak value, it was possible to extend the frequency range from 0.5 to 8.0 Hz, because there was no restriction due to limiting radiation dosage. The movements of the wire pins could be analyzed by means of redrawing and subsequent measuring on a table. This method allowed analysis of the maximal displacement of the individual vertebral bodies in the vertical and horizontal directions, in addition to the motion patterns of the vertebral bodies and the maximal bending vibration in the sagittal plane.

The relative changes in vertical vibration amplitude compared to the induction site at the seat are presented for the cervical and lumbar areas in

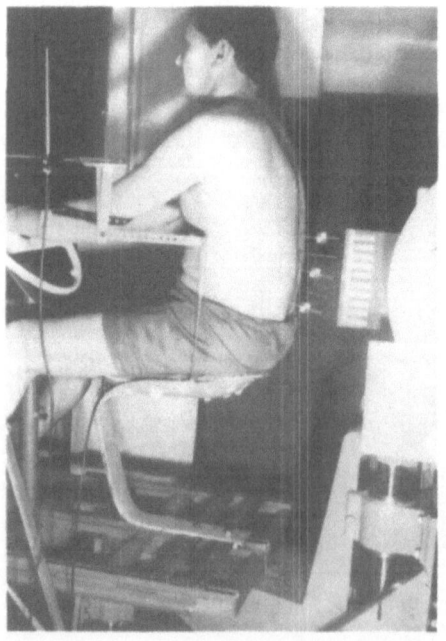

Fig. 17. Subject on the vibration simulator with vertebrae T12 and L 1 – 4 indicated by wire pin markers

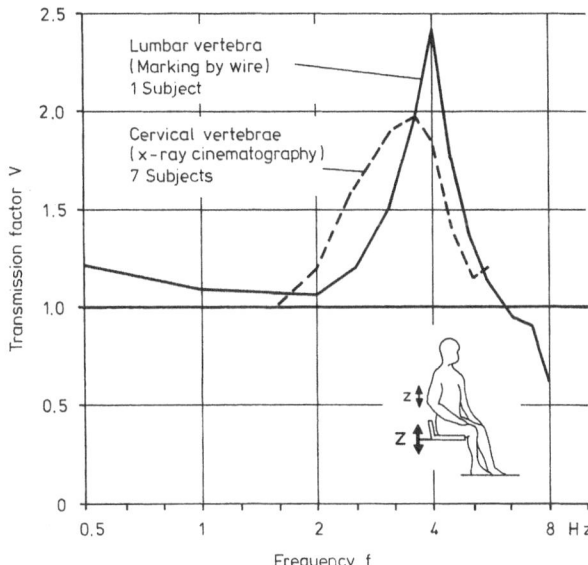

Fig. 18. Frequency-dependent vibration transmission from the seat to the lumbar spin (*LS*) and cervical spine (*CS*) in the vertical direction z (Dupuis 1969)

Fig. 18. The cervical spine shows, with increasing frequency, an increase in vibration amplitude with a resonance peak at 3.5 Hz, which then again falls. These results correlate well with the acceleration measurements (Dupuis 1960; Dieckmann 1957) that were done in the shoulder-neck area. The films from the cineradiograph also demonstrate the special sensitivity of the cervical area in the frequency range of 2.5 and 5.5 Hz, at which the compression and extension of this segment is very pronounced. The resulting deformation of the invertebral disk is clearly recognizable. However, no quantitative analysis was made of this optic impression while watching the cineradiograph.

The lumbar spine has a resonance between 2 and 6 Hz with a maximum at 4 Hz, at which the vibration amplification increases to 240% (Fig. 18). This shows a good correlation with the experiments of Lange and Coermann (1965), who had established the greatest vertical shifts between two neighboring lumbar vertebrae at frequencies of 4−5 Hz, with an amplitude in the order of magnitude of 0.5 mm (3.3 m/s² vibration stimulation). Above 6 Hz, a decrease in the vibration transfer took place.

In spite of pure vertical vibration excitation through the seat, the lumbar vertebral column also moves in the horizontal direction, x, as is shown in the data presented in Fig. 19. The L4 exhibits the greatest displacement, which decreases L2 to T12.

This behavior can be explained, as the unsupported body, in order to maintain equilibrium, tries to balance movement in a sagittal direction. This motion tendency has been detected previously in tractor drivers by means of light-point motion pictures (Dupuis 1960).

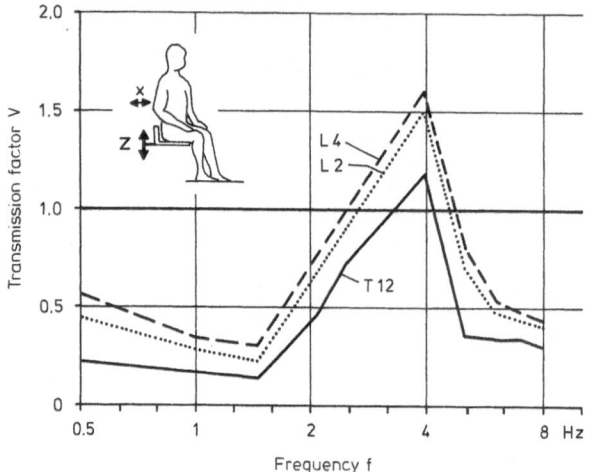

Fig. 19. Frequency-dependent vibration transmission from the seat to lumbar vertebrae L4 and L2 and thoracic vertebra T12 during vibration in the vertical direction z (vibration component x on the vertebral segment; Christ and Dupuis (1969)

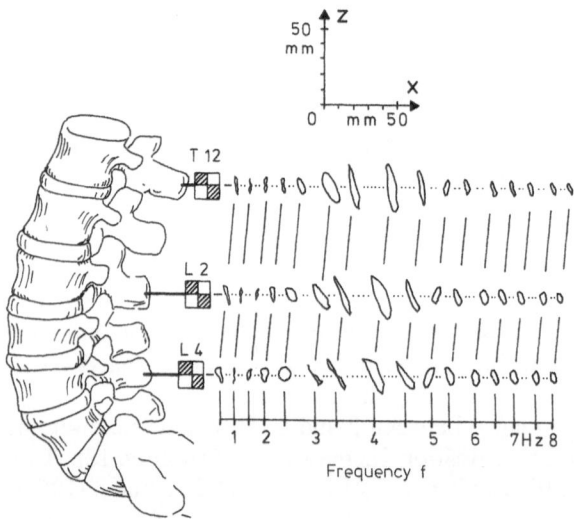

Fig. 20. Movement forms of the wire pin markers for three vertebrae during excitation from vertical sinusoidal vibration with 5-mm displacement amplitude (peak value). Christ and Dupuis (1966)

Representation of the total motion pattern in the sagittal plane (Fig. 20) confirms that at 1.5 Hz the lumbar part of the spine is relatively motionless in both directions. It can be clearly seen that the vertebrae, corresponding to the motion of the lumbar vertebrae segments, respond to vertical input with a vibration with the main direction, zzx, that is to say, from the vertical forward to 35°. Above all, this pertains to the frequency range > 3 Hz, whereas from 0.5 to 2.0 Hz the movement is mainly vertical. The vibration patterns change − independent of their main direction – from almost straight lines over ellipses to circular movements.

Angle changes also occur during the movement of vertebral bodies. These changes were each measured for the various vibration periods as maximal rotational displacement in single-picture data. They are shown in Fig. 21 in relation to the stimulation frequency. The greatest angle movement generally occurs at T 12. That can be explained by the fact that the total mobility increases of L 4 over L 2 to T 12 because it is added to the mobility of the underlying segments. Only below 3.3 Hz does the L 2 show a greater bending vibration than T 12, which is indicative of reflex movement of neighboring segments. This can also be seen on film, but was not quantitatively evaluated. The least bending vibration is at L 4.

For all three vertebrae in Fig. 21, the strongest angle changes occur in the resonance region of the whole body at 3 – 5 Hz, at which the amplitudes are almost doubled. Below 2 Hz, however, the bending angle increases slightly, like the vibration displacement.

Using the same experimental methods as Christ and Dupuis (1966), experiments were carried out using stress from random vibration. The results showed that an optimally designed seat-spring system, as compared to a simple leaf-spring seat, can reduce vertical vibration displacement of the cervical part of the spine by 49% and of the lumbar spine by 53%.

In addition to these measurements in vivo, it appears necessary to acquire knowledge on the vibration forces and on the mechanism that can destroy the spinal column. Knowledge such as the breaking strength of sections of the spi-

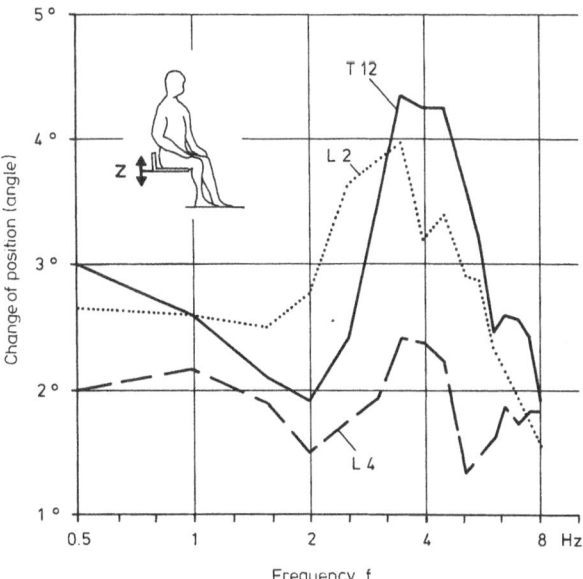

Fig. 21. Rotational vibration of three vertebral bodies in the sagittal plane in the vertical direction z during sinusoidal vibration with 5-mm displacement amplitude (peak value). Christ and Dupuis (1966)

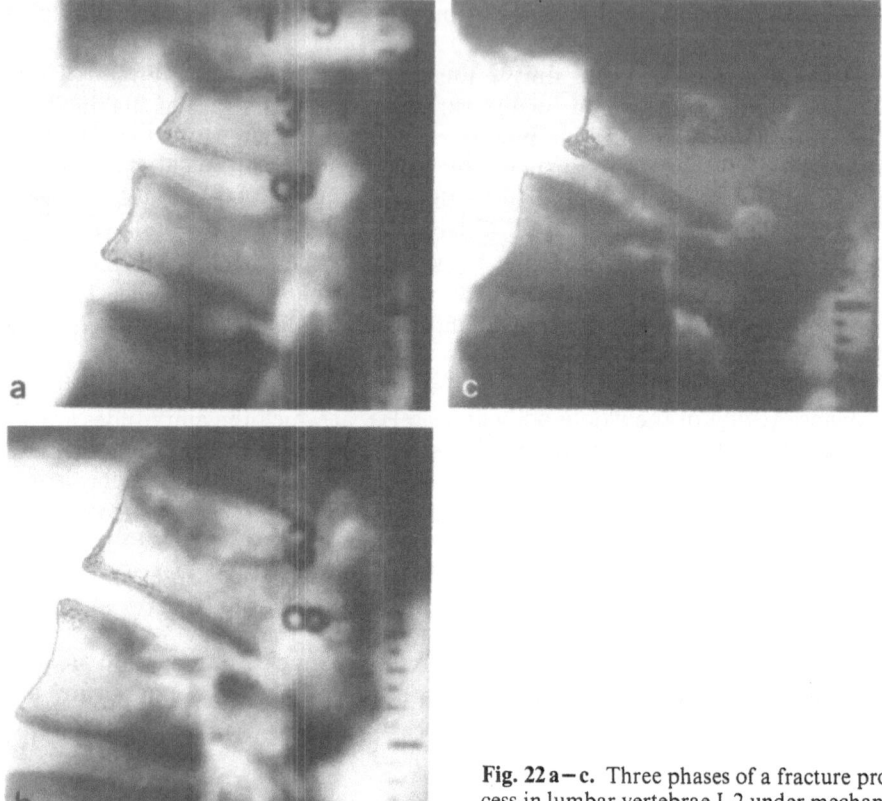

Fig. 22 a–c. Three phases of a fracture process in lumbar vertebrae L 2 under mechanical pressure (Christ and Dupuis 1966)

nal column, of individual vertebral bodies, or of the intervertebral disks is necessary in order to be able to estimate short-term tolerance during vibration stress (Kazarian 1972; Kazarian and Graves 1977). Brown et al. (1957) and Hirsch (1965) have also carried out such static experiments, and Evans and Lissner (1959) have tried to determine axial deformation. In addition, there have been many clinical observations on the intervertebral disks. However, to our knowledge little research has been done on dynamic behavior under extreme vibration exposure. To make a contribution to this area, we tried to observe the breaking process of an excised vertebral body by means of roentgenocinematographic imaging, with simultaneous recording of the applied forces.

For this purpose fracture experiments were conducted on a lumbar vertebral section of a corpse (Christ and Dupuis 1966). The upper and lower segments of the vertebral column were fixed with a vice; then the spine was exposed to increasing pressure with the piston of the hydraulic cylinder of the "hydropulse" equipment. The pressure was measured by a pressure gauge, which registered continuously. The breaking process itself was continuously observed on the television screen and recorded on film.

In Fig. 22a–c, the fracture process is represented in selected X-ray film pictures in three phases. It can clearly be seen that the upper part of the vertebra is the first to break. Finally, the vertebral arch also shears off with part of the dorsal ridge. Clinical experience indicates that these upper-plate breaks are typical fracture forms (Böhler 1953). The maximum force applied, which led to the fracture, was 8820 N.

The knowledge represented in this section on the biodynamic vibration behavior of the spine is also important in connection with the interpretation of chronic health injuries to the spinal column under vibration stress (see section "Diseases of the Spinal Column," p. 91).

5.2.7 Vibration Behavior of Internal Organs

In previous sections, it was shown that resonance vibration occurs in the whole trunk and in the spinal column during vibration stress in the standing and sitting position. Resonance always means that there is especially great biodynamic strain on the body tissue involved. It must therefore be assumed that the internal organs of the trunk also have resonances, depending on the excitation frequency, and that the natural frequency and amplification depend on the mass of the organ, the spring rate, and damping of the tissues (tendons, ligaments, connective tissue, vessels) with which the organ is connected to the vibrating structure.

In his survey paper dealing with the effects of mechanical vibration on the organism, Rublack (1978) cites approximately 25 original publications in which research in this field has been reported. However, his brief comments on each of his citations do not give any details regarding each experiment.

It is understandably difficult to carry out experiments on the vibration behavior of the internal organs, inasmuch as each organ cannot be excited to vibrate on an individual basis and cannot be measured noninvasively. Therefore, it is necessary to rely upon indirect measurements and observations.

Early experiments of this kind include those carried out mainly on animals in the Aerospace Medical Research Laboratories in Ohio. In these experiments radiographic and physiological measurement procedures were used to measure respiration. The thorax and abdomen were the objects of these experiments.

With the help of *roentgenkymography*, Nickerson and Coermann (1962) observed the internal structures of the thorax and abdomen in anesthetized dogs that were exposed to vibration stress of up to 10 Hz in the z-direction parallel to the vertebral column. They found that both body regions vibrate like pure masses, with resonance frequencies of 3–4 Hz and damping factors of 0.2–0.25. The observation that the phase angle reached 180° in some cases led them to conclude that the system cannot be understood in terms of a simple damped oscillator (one degree of freedom), but as a coupled vibrating system (two or more degrees of freedom).

They compared the results of animal experiments with human data, which had shown that the visceral mass of the thorax and abdomen vibrates with a resonance frequency of 3–5 Hz and with damping factors of 0.2–0.25. The variation of the resonance peak, which was found to be greater in the abdominal region than in the thoracic area, were explained by the varying tension in the abdominal wall. Occasionally, another maximum was observed in the frequency range of about 7–10 Hz. When resonance exists, the authors assume that there is considerable strain in the tissues. Vibration-caused changes in respiratory function are taken as an indication of resonance in lung tissue (see section "Respiratory Function," p. 57).

Using a research method that was somewhat different, Nickerson and Drazic (1966) observed the vibration behavior of the abdominal viscera of anesthetized dogs during vibration stimulation in the three axes, x, y, and z, while the vertebral column was immobilized. Several months before the beginning of the experiment tantalum metal plates had been operatively implanted into different body regions and, under the given research conditions, observed by roentgenkymography. The results from these experiments are shown in Table 5.

The variation in resonance frequency is the most striking for the x- and y-direction compared to the z-direction. Although in the caudiocranial direction, z, all internal organs showed resonance between 4 and 5 Hz, as established earlier, it was determined that the resonance frequency was considerably higher in the dorsoventral direction, x, and in the lateral direction, y, with an overall average of about 9 Hz. The authors attribute this phenomenon to the relatively low mobility of the viscera in the x- and y-directions because of (partial) mobility limitations by the ribs, vertebral column, and pelvis.

As these experiments were carried out on dogs, transfer of the results to humans is subject to certain scaling laws (von Gierke 1968, 1971) and is possible only within certain limitations. For man, somewhat lower resonance frequencies are to be expected, which would correspond to the greater mass of these body parts in man.

Table 5. Resonance frequencies f_0 and damping factors D of different internal organs in dogs under vibration stress in three directions. (Nickerson and Drazic 1966)

Organ	Dorsoventral x		Lateral y		Caudocranial z	
	f_0	D	f_0	D	f_0	D
Thorax-aorta	7.7	0.31	9.9	0.33	4.3	0.36
Heart apex	9.8	0.40	8.2	0.46	4.8	0.38
Ribs	8.6	0.30	9.5	0.51	4.3	0.31
Diaphragm	7.9	0.26	8.6	0.30	5.0	0.41
Stomach	11.0	0.58	10.4	0.47	4.2	0.21
Intestines	7.9	0.39	10.1	0.40	4.2	0.21
Rectum	8.0	0.44	8.6	0.34	4.4	0.36
Bladder	10.3	0.46	8.1	0.20	4.5	0.42
Right kidney	8.6	0.26	10.1	0.40	4.2	0.23
Mean value $\bar{x}$	8.9	0.38	9.3	0.38	4.4	0.32

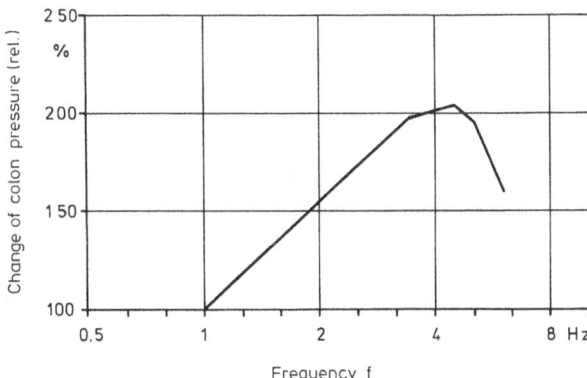

Fig. 23. Frequency-dependent change in colon pressure under vibration (White et al. 1963)

For measurements in human beings, Bruner (1960) developed a method for obtaining data on the interior pressure in the gastrointestinal region and found that if the pressure measured 90 mm Hg, it was generally connected with pain. At approximately the same time, Ziegenrücker and Magid (1959) found that test subjects complained of pain in the stomach-intestinal region when they were exposed to vibration on a shaking platform at a frequency of 4–8 Hz and acceleration amplitudes of 20 m/s². The first conclusion drawn from this information was that the interior pressure of the colon must not reach 90 mm Hg.

White et al. (1963) based an experiment on this assumption. Introducing an air balloon into the colon, which was connected to the outside by small tubing, they continuously recorded the pressure by means of an inductive transducer. The colon pressure was measured in eight subjects. The results of this experiment showed an increase in colon pressure as the frequency increases from 1 to 4.5 Hz, and above that there was a decrease (Fig. 23). The interior pressure was also measured in the stomach of an anesthetized rhesus monkey under the influence of mechanical vibration for the range 1–20 Hz.

Nutrition physiologists, orthopedists, radiologists, and electronic experts collaborated in further experiments, in which the vibration behavior of the stomach was determined when under the influence of sinusoidal and random vibration (Dupuis and Christ 1966a).

The roentgenocineradiography equipment already mentioned in the section "Vibration Behavior of the Spinal Column" (see p. 31) was used in this experimental procedure. Five male students 20–26 years of age volunteered to serve as test subjects, and all were first examined to be sure that there were no abnormal gastrointestinal findings. Besides strict observance of radiation protection rules, a Mavig gonad-protection device was used in all cases and, as far as possible, protection through lead rubber sheets was provided. Radiation scatter was reduced by narrowing the projection surface. The filming duration and radiation dosage were reduced as much as possible.

Immediately before the experiment started, four test subjects received a meal of solid "normal" food (105 g beef, 60 g noodles, 20 g fat), which was enriched with 30 g Unibaryt C as a contrast medium. In this way, it was possible to fill the stomach physiologically, on the one hand, and to represent its contents on the other by means of the contrast medium. Instead of solid food, the fifth test subject was given a liquid meal of 0.45 l beer enriched by 25 g Unibaryt C.

The results of the experiments with vertical sinusoidal vibration, with a peak value of vibration displacement of 8 mm at the seat, at a frequency range of 0.5−6.0 Hz, are shown in Fig. 24. The mean value curve for the five subjects (four subjects with normal diet and one with liquids alone) shows an increase in vibration displacement opposite the induction site (seat), which almost doubled at 4.5 Hz (Fig. 24). This vibration behavior corresponds to the results by White et al. (1963) with regard to the interior pressure in the large intestine (Fig. 23). Although no pressure measurements have been carried out in the stomach of man, animal experiments indicate that the course of stomach pressure can also be expected to show the same dependency on frequency.

The individual resonance curves of the stomach for four subjects on a solid diet show that there are considerable differences in the amplification. These differences may be explained to a certain degree by the various stomach forms, but above all by the differences in tone of the stomach and abdominal musculature, resulting in varying degrees of damping of the spring-mass system of the stomach. With regard to the frequency range of the resonance peaks, there is a close conformity in the area of 4−5 Hz.

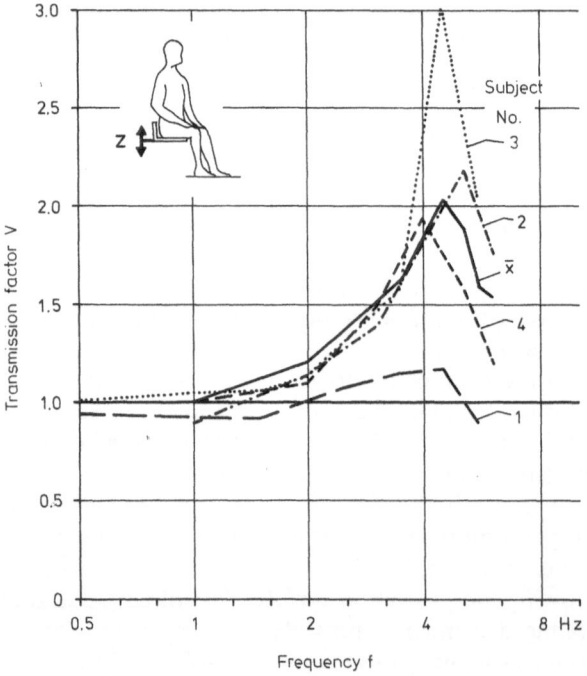

Fig. 24. Frequency-dependent vibration displacement on the stomach in the seated position during vibration in the vertical direction z (Christ and Dupuis 1966 a)

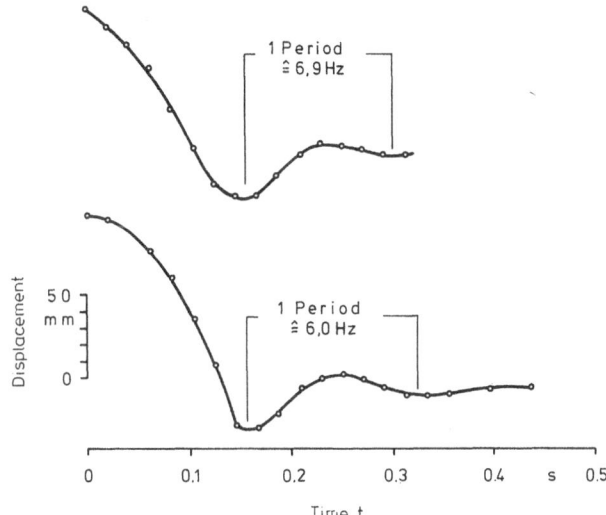

Fig. 25. Damped oscillation curves of the stomach in two subjects, as evaluated in X-ray films (Christ and Dupuis 1963)

Similar conditions are to be expected if the stomach is not filled with solid food but with liquids. A research series carried out for comparison resulted in a similar transmission curve between 0.5 and 3.5 Hz. However, after a short time the contrast fluid had left the stomach so completely that an image of the stomach could no longer be obtained on the X-ray film in order to carry out further experiments at frequencies of > 3.5 Hz. Even though the series could not be continued for this reason, a resonance comparable to the solid food case can be expected on the basis of the very similar vibration behavior up to 3.5 Hz.

Regarding the question of the natural frequency and damping behavior of the stomach, vibration measurements were also carried out in two subjects. When the whole body was allowed to drop on a solid seat, X-ray analysis showed that the vibration event of the stomach was strongly damped (Fig. 25). From this information it was possible to deduce the natural frequency, f_0, of the stomach in these two subjects at 6.0 – 6.9 Hz. It was not possible to formulate this finding very precisely, however, since the stomach vibrated only for a short time afterward, and the vibration behavior may be influenced by the tension of the musculature of the abdominal wall, as well as of the diaphragm.

If the greatest vibration displacement of the stomach is found between 4 and 5 Hz, then this is in agreement with the results in dogs by Nickerson and Drazic (1966; Table 5) and is caused by the resonance of the whole body in the z-direction. It is practically impossible to stimulate the natural frequency of the stomach alone to natural vibration without exciting the whole body resonance.

The vibration behavior of the stomach under simulated random tractor vibration was further investigated using two different types of seat. Evaluation of single-frame cineradiographic shots has proven that with the use of a seat with parallelogram suspension, appropriate damping and adjustment capability for body weight, vertical vibration displacement can be reduced, as compared to the use of a simple leaf-spring seat (Dupuis and Christ 1966a). The reduction achieved at the base of the stomach was 43.7% and at the height of the fluid level in the stomach, 49.2%.

The use of abdominal support belts (waist belts) is occasionally recommended for motorcycle riders and sometimes is also included in the regulations for military troops and police units. It is assumed that such belts provide protection of the internal organs against mechanical vibration. To clarify whether these assumptions are correct, two subjects were exposed to random vibration, first with and then without a waist belt, and X-ray films were taken of the stomach.

The results provide no proof that such belts have a vibration-reducing effect. On the contrary, the mean values of both subjects − taking individual differences into account − indicate that the waistbelt might bring about a certain amplification of the vibration displacement. [According to experiments by Coermann (1963), the use of a waist belt also increases the mechanical impedance.] It is possible that fixation using a waist belt may reduce the potential horizontal movement of the viscera and thus intensify vertical movement. Freytag and Puzyna (1953), who conducted practical experiments on 17 tractor drivers, have also expressed doubt as to the usefulness of waist belts − at any rate with regard to reducing vibration of the stomach.

Although Dupuis and Christ (1966a) have conducted experimental trials to answer the question of how the degree of fullness of the stomach can influence vibration behavior, to date there have been no definite results because the food that had gone into the intestines was replaced immediately by secretions and a half or completely empty stomach cannot be visualized on X-ray film.

The findings represented here on the vibration behavior of the stomach are also meaningful in connection with the interpretation of epidemiological findings (see section "Digestive System Diseases," p. 114).

5.2.8 Vibration Behavior of the Eye

Impairment of visual performance under the effect of mechanical vibration can often be traced back to blurring of the picture on the retina. Knowledge of the frequency-dependent vibration behavior of the bulb in the eye socket is therefore very useful in evaluating the influence of vibration stress on visual performance.

In previous well-known experiments on the reduction of visual perception caused by vibration, resonance of the eyeball is often implicated and sometimes data on resonance frequency are included in the reports. However, in most cases these statements are not based on experimental evidence but are usually hypothetical assumptions. Only a few experiments have been carried out on the vibration behavior of the bulb.

In 1938, Coermann had already carried out numerous experiments on the physiological effect of mechanical vibration. In this connection, visual acuity was also determined during vibration stress at frequencies of between 20 and 100 Hz. Under constant acceleration, especially strong discontinuities in visual acuity were established at 25−40 Hz and 60−90 Hz. These results were interpreted by Coermann as being caused by the natural vibration frequency of the eyeball. Since then, many publications on this problem have dealt with Coermann's hypothesis, but most have been unable to provide any further experimental support.

In field experiments in which the biodynamic transmission of vibration was investigated from the induction site to the orbit in 31 types of human locomotion, Dupuis et al. (1975) established high vibration acceleration in the frequency range of 20 Hz while the subject was waterskiing over water with slight waves. The authors attributed the blurred vision of the subjects as possible eyeball resonance.

Ohlbaum (1976) critically looked at the results of two other American experiments. In experiments with anesthetized dogs, Nickerson et al. (1963) had established a sharp resonance peak at about 40 Hz for the bulbs of these animals. Since the input acceleration was measured at the teeth but not immediately next to the eyes, Ohlbaum did not accept the results as demonstration of the natural vibration frequency of the eyeball. He also doubted the correctness of the results of an experiment by Lee and King (1971) who had found a nontypical resonance behavior for the bulbs in human experiments using complicated indirect methods. This is probably correct, as the subjects, for technical reasons, had to wear spectacle frames and the natural vibration frequency of the frames may falsify the results.

In 15 subjects, Ohlbaum found experimentally a mean resonance peak at 18 Hz, using photographic methods utilizing the cornea reflex. In order to record the acceleration, however, again spectacle frames were used, which can adversely affect the measuring values. In conclusion, one must agree with Shoenberger (1972), who stated that the determination of natural vibration behavior of the bulbs is methodologically difficult and, for that reason, the results attained so far should not be used uncritically.

To clarify this question, further corresponding experiments have been conducted using other methods on animals and subjects (Dupuis and Hartung 1979, 1980a, b; Hartung 1983).

Dupuis (1977) and Dupuis and Hartung (1979) conducted preliminary experiments, using swine skulls, approximately 1 h after the death of the animal. They attached miniature accelerometers (weight 0.5 g, size $8 \times 3.6 \times 3.6$ mm) in the vertical measuring direction on the cornea (by sewing) as well as on the cranium (by screwing), which was stripped of skin and muscle tissue. The acceleration of the eyeball and of the skull was measured under vertical vibration stimulus, with constant acceleration in two steps ($a_{eff} = 2.0$ and 4.0 m/s²) in a frequency range of $1-400$ Hz.

These animal experiments have shown that such a procedure is basically possible and can also be used on animals in vivo. Therefore, major experiments were carried out in vivo on two primates of the Macaca species (one *Macaca mulatta* and one *Macaca cynomolgus*). Attachment of the accelerometers was carried out in the same way as in the preliminary experiments.

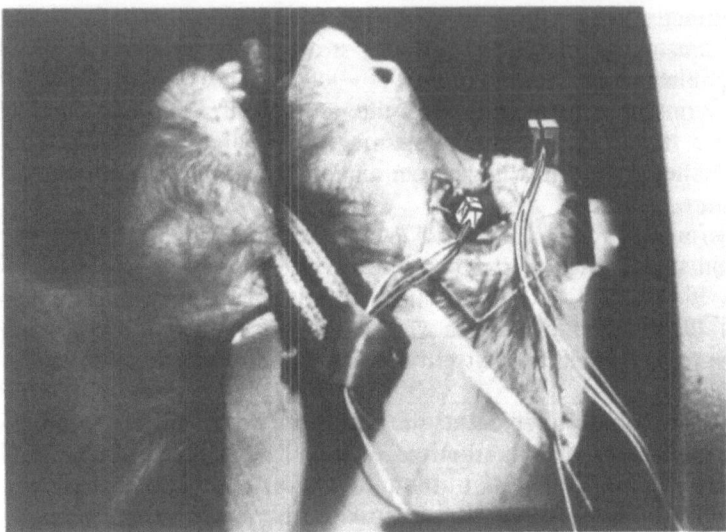

Fig. 26. Vibration stimulation of the head of a primate with miniature accelerometers firmly fastened to the cornea and skull (Dupuis and Hartung 1979, 1980a)

The animals were completely anesthetized with nembutal so that there was no influence on the resting tone of the musculature. Because of the narcosis the animals had to remain in a reclining position while being stimulated by horizontal vibration. The vibration stimulus was therefore aligned caudocranially (Fig. 26).

In contrast to all of the above-mentioned experiments of other authors, the vibration was induced immediately into the cranium in order to have accurate vibration exposure, as defined by frequency and acceleration at the skull (os frontale).

The results of the experiments with primates are represented as mean values in Fig. 27. An increase in vibration in the bulbs was found at between 12.5 and 40 Hz and > 63 Hz. The resonance maximum can be found between 25 and 31.5 Hz, as the rising curve above 63 Hz cannot be explained by resonance of the eyeball. Instead, it has to be assumed that resonance of the lightweight accelerometer was excited since the fastening threads could not be tightened at will.

Since in human experiments the attachment of plane mirrors, fixed contact lenses, or miniature accelerometers is not possible, experiments have been conducted using the principle of reflecting a light beam by the cornea (Doden 1976; Dupuis and Hartung 1980a, b). In order to do this, a reflective point on the cornea of the subject was produced in a darkened laboratory with a light source. For a comparative light point, a miniature lamp was fixed directly above the eye on the forehead, stabilized by a headband and a bite bar (Fig. 28).

A multipoint X-Y-tracker television system was used for the measuring procedure. The detailed picture (eyes and forehead lamp) was continuously taken by a camera with a telephoto lens, mounted on a tripod. The image intensifier, equipped with a scope, made the picture visible, fixated both light points (cornea reflection and forehead lamp), and followed their movements in two dimensions in the vertical frontal plane (X-Z tracker). The movements corresponding to the vibrations of both light points were converted into electric voltages and recorded on a cathode-ray oscilloscope and a direct-recording instrument.

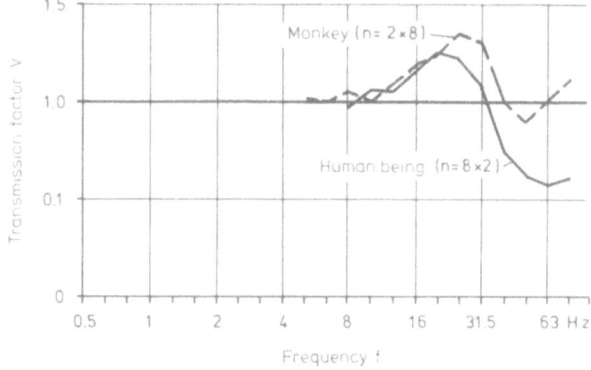

Fig. 27. Frequency-dependent vibration displacement from the skull to the cornea of the eyeball during vibration in the z direction (Dupuis and Hartung 1979, 1980 a, b)

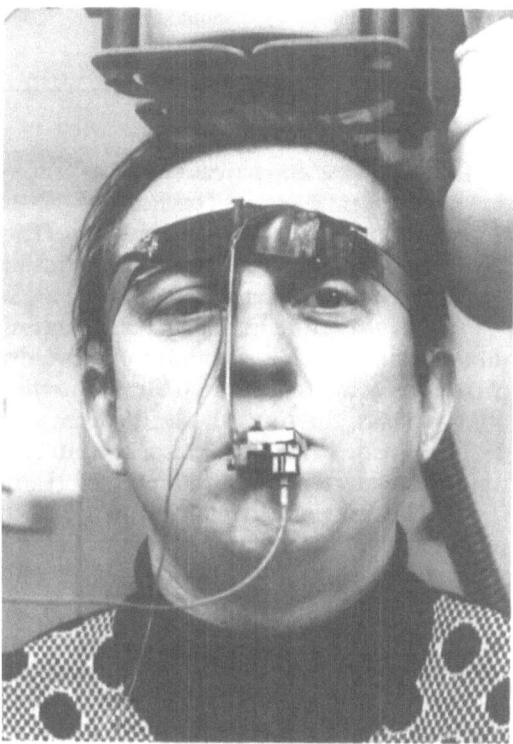

Fig. 28. Light points on the cornea and forehead, with the accelerometer fixed on the bite bar (Dupuis and Hartung 1980 a, b)

Table 6. Resonance frequencies f_0 of different human body parts under vibration with various body postures and vibration directions

Body posture	Body part	Vibration direction (see Fig. 1)	Range of resonance frequency
Reclining	Foot	x	16 –31 Hz
	Knee	x	4 – 8 Hz
	Abdomen	x	4 – 8 Hz
	Chest	x	6 –12 Hz
	Skull	x	50 –70 Hz
	Foot	y	0.8– 3 Hz
	Abdomen	y	0.8– 4 Hz
	Head	y	0.6– 4 Hz
	Foot	z	1 – 3 Hz
	Abdomen	z	1.5– 6 Hz
	Head	z	1 – 4 Hz
Standing	Knee	x	1 – 3 Hz
	Shoulder	x	1 – 2 Hz
	Head	x	1 – 2 Hz
	Whole body	z	4 – 7 Hz
Sitting	Trunk	z	3 – 6 Hz
	Chest	z	4 – 6 Hz
	Spinal column	z	3 – 5 Hz
	Shoulder	z	2 – 6 Hz
	Stomach	z	4 – 5 (7) Hz
	Eyes	z	20 –25 Hz

Eight institute employees served as subjects and each completed two runs of each experimental series. The head of the subject was exposed to direct vibration by means of a vibration simulator. The acceleration intensity was selected by the subject who chose the intensity just below tolerance. Each vibration was measured by an accelerometer fixed on the bite bar (Fig. 28).

The results of the human experiments using the TV X-Y-tracker system are presented in Fig. 27 in the form of a mean value curve for eight subjects, 2 experiments each (16 data sets). Similar to the animal data obtained by using another method, there was a clear increase of bulb vibration between 12.5 and 31.5 Hz, with a maximal resonance at 20.25 Hz. The greatest amplification transmission was V = 1.23 (Dupuis and Hartung 1978).

Although the mechanical properties of the vitreous body of swine and primates are comparable to the human eye, transfer of the results from these animal experiments to the physiological conditions of man is not possible without certain limitations.

The argument that the use of narcosis resulted in nonphysiological conditions can be refuted inasmuch as the narcotic selected permits the resting tone of the eye muscles to be uninfluenced. Nevertheless, it cannot be excluded that the horizontal body position compared to the vertical position can have an influence on the results. It can further be argued that the dimensions and weight of the vitreous bodies of subhuman primates are smaller than those of man. The lesser mass of the subhuman primate bulbs (4.1 versus 7.5 g in man) should, however, result in a somewhat higher effect in natural frequencies, if one assumes similar spring constants for the tissues surrounding the vitreous body.

The experimental results support this trend: the maximum natural frequency of the eyeball of small primates is $25-31.5$ Hz; that of the human bulb, however, is $20-25$ Hz, with a maximal transmissibility factor of 1.23. Since for the human eye a transmissibility factor of > 1.0 has been observed over the total frequency range $10-31.5$ Hz (Fig. 27), there should also be a decrease in optical perception in this frequency range. Experiments by Ohlbaum (1976) have established a resonance range predominantly between 12.5 and 25 Hz, with a maximum at 18 Hz and a transmissibility factor of 1.33.

In these experiments it cannot be differentiated whether the established resonance movements of the bulbs correspond exclusively to translational movement, rotational movement, or to a combination thereof. We assume it is the latter, but this question is irrelevant since translational as well as rotational eye movements caused by vibration impair visual perception. Thus, when vibration excites the eyeballs in other directions, particularly in the horizontal transversal axis, the possibility of resonance with its resulting influences on the visual performance cannot be excluded.

5.2.9 Review of the Resonance Frequencies of Various Parts of the Body

A summary of the most important areas of resonance frequency for the various parts of the body are shown in Table 6, as based on the experimental results presented. It may be seen that under different conditions (body posture, vibration direction), the parts of the body are in resonance at varying frequencies. In general, low frequencies prevail. No further knowledge on the natural vibration behavior of the human internal organs is available. Table 5 gives indications for resonance frequencies of some internal organs, as established in dogs.

When constructing technical equipment, it should be taken into consideration that, if possible, those vibration frequencies at which body resonance can be expected should be avoided, particularly when vibration transmission to man cannot be avoided.

5.3 Changes in Physiological Functions

In the discussion on the biodynamic vibration behavior of different parts of the body, it has already been pointed out that strong vibration stress, e.g., that leading to resonance vibration of the organ systems, can cause physiological[2] reactions on the basis of high mechanical stress on body tissue alone.

Such reactions can affect muscle function, the respiratory and circulatory systems, the vegetative nervous system, sense perception, and biochemical changes. Thus, in a study of the literature, Seidel (1975) has systematically compiled physiological reactions to whole-body vibration in the vertical direction.

Symptomatic complaints in various organ areas, dependent on the stimulating frequency, can indicate vibration-induced physiological reactions, such as those reported by Magid and Coermann (1960). Seated subjects were exposed to vibration of varying frequencies from 1 to 20 Hz. The amplitudes were slowly increased until the subjects felt that the acceleration was so strong that they feared further increase might result in damage. Afterwards the subjects described the kind of complaints and pointed out the organ regions where the complaints originated or where there was pain. Figure 29 presents these symptomatic complaints in conjunction with vibration frequency.

The preciseness of such subjective findings may be small because of the difficulty of obtaining quantitative information on certain symptoms and reactions

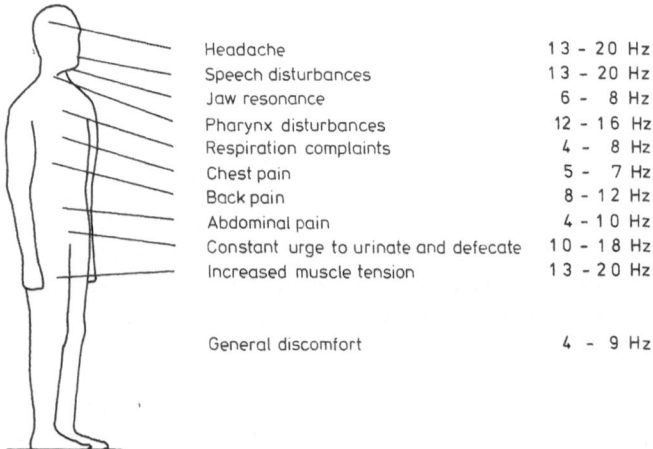

Headache	13 - 20 Hz
Speech disturbances	13 - 20 Hz
Jaw resonance	6 - 8 Hz
Pharynx disturbances	12 - 16 Hz
Respiration complaints	4 - 8 Hz
Chest pain	5 - 7 Hz
Back pain	8 - 12 Hz
Abdominal pain	4 - 10 Hz
Constant urge to urinate and defecate	10 - 18 Hz
Increased muscle tension	13 - 20 Hz
General discomfort	4 - 9 Hz

Fig. 29. Complaints in various organ regions in relation to stimulating vibration frequencies (Magid and Coermann 1960)

---

[2] The concept "physiological" is not used here or in the text to follow in the evaluating sense, such as "physiological normal," but instead in the sense of changes in the physiological parameters; in general, it is open as to whether these changes lie in the normal region of physiological scatter or in the pathophysiological region

and because of the considerable interindividual deviations. The information is insufficient to establish strain scales or dose-effect relationships (Bobbert 1967). Therefore, it is desirable to describe the effect of mechanical vibration on the human body not only by means of biodynamic reactions and subjective impairment, but also by changes of physiological parameters.

Physiological reactions to mechanical vibration can be but are not necessarily perceived by the people who are exposed. At first, the reactions are acute and therefore can sometimes be measured. Sometimes they are not immediately noticeable at the beginning of the mechanical stress but are perceived after 1 or 2 h, such as the symptoms in kinetosis (see section on "Kinetosis," p. 66). In this case, it is called "sea sickness," but is by no means chronic. Reactions of a physiological type can have chronic pathophysiological consequences, however, if they occur continuously and over a long period (Freund and Dupuis 1974). More will be said on this matter in the section entitled "Chronic Effects of Whole-Body Vibration" (see p. 87).

5.3.1 Muscle Activity

From the knowledge available on the mechanical vibration behavior of individual parts of the body, it must follow that man will always try, consciously or unconsciously, to counteract the vibration if it is annoying. If he reacts to vibration with active body movement, he will have to spend muscular energy. This also applies to more passive behavior, "involuntary shaking," as he will try as a reflex reaction to maintain equilibrium by muscle action.

If man is exposed to sinusoidal or periodic vibration, with which he has become acquainted through its frequent occurrence, he will respond partly with active, dynamic muscular action. The movement behavior is compensation for vibration in order to make it less annoying. To a certain degree static muscular action is taking place.

However, in the case of random vibration, there is hardly any dynamic muscular reaction. Instead, man tenses up with extensive static muscular action. This "pretension" of the most important muscle groups serves as a quick reaction, but occasionally the purpose is simply to change the natural frequency and the damping rate of the body, so that the resonance effects can be avoided or reduced.

Dupuis et al. (1972) have clarified this muscular behavior experimentally: electromyograms of the m. erector spinae were taken using surface electrodes during the influence of sinusoidal and random vibration in the horizontal x-direction on seated subjects (Fig. 30). The electromyogram shows that when the muscle is under sinusoidal stress of low frequency, it answers alternately with tension and relaxation phases that are synchronous to the vibration. Apparently this results in the development of a "behavior pattern" that permits the musculature to react economically and efficiently with contraction and relaxation (shown by alternating large and small amplitudes). This is different, however, when the stress is from random vibration. Because of the lack of a

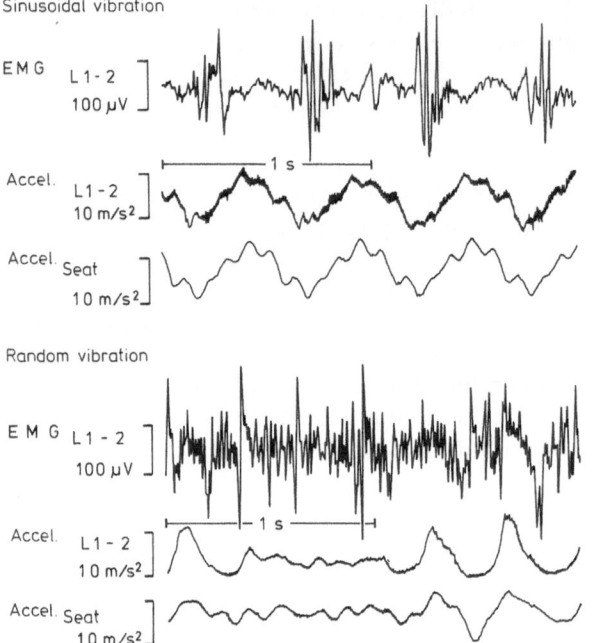

Fig. 30. Electromyograms (*EMG*) of the m. erector spinae of seated subjects under the influence of sinusoidal (*above*) and random vibration (*below*) in the horizontal direction x (Dupuis et al. 1972)

clear "behavior pattern," the result is static muscle contraction without recovery phases. These relationships also explain why man reacts more sensitively to random vibration than to sinusoidal vibration when the rms value is the same (Dupuis 1975 b).

Examples of defensive muscular reactions have been given by Sassor and Krause (1966). The contraction of the abdominal wall musculature is increased under certain stimulus frequencies in order to keep the movement of the abdominal viscera slight. With this musculature contraction, an increase in the natural frequency of the abdominal viscera occurs, resulting in more favorable tuning.

With regard to vibration transmission, contraction of the musculature is also important for the different body postures, as is shown in the section "Biodynamic Reaction on Vibration" (see p. 14 ff).

Some studies that have been carried out on the reaction of the musculature when under vibration stress have not provided any quantitative data. Either the electromyographic activity was not correlated to the physical vibration parameters (Borscevskij et al. 1963), or only general qualitative data were obtained (Butkovskaja and Kadyskina 1970).

Other studies, on the other hand, have permitted the establishment of stress--reaction relationships: Guignard and Travers (1959) have conducted electromyographic measurements on seated subjects during whole-body vibration.

In the musculature of the lower extremities they were able to show electric activity that increased with increasing vibration stress. Schäfer (1977) also found an increase in the activity of the musculature in the region of the lumbar vertebral column during simulated vehicle vibration with increasing vibration intensity.

Berthoz and Wisner (1968) and Berthoz et al. (1972) have also carried out extensive electromyographic measurements in connection with the biodynamic vibration behavior of the seated person during vertical vibration stress at low frequencies. These measurements were carried out on the paravertebral muscles in the region of three vertebral segments: C4, T10, and L1. Synchronous muscular reactions were observed during the vibration process. The muscle activity increased in the region of the resonance frequency of the trunk (3.9 − 4.9 Hz), as well as with increasing exciting vibration displacement amplitude. In the neck musculature, intensified potentials were shown in the electromyogram of up to 6 Hz versus 2 − 3 Hz.

Using seated subjects, Cursiter and Harding (1974) also found electric activity in the m. trapezius and the remaining neck musculature acting synchronously with the alternating vibration event.

With the exception of the electromyographic measurements presented in Fig. 30, the experiments described so far deal with the effect of vertical vibration in the z-direction and with seated subjects. Weiner (1973) has carried out electromyographic measurements on the extensors of the back, using ten subjects in an upright position without a backrest, who were exposed to sinusoidal vibration in the horizontal x-direction (dorsoventral). Using statistical analysis of variance, he was able to show a correlation between electric muscle activity and biodynamic vibration behavior. With increasing resonance of the trunk in the dorsoventral direction, the dynamic muscular activity of the m. erector spinac is increased. This is the case at 3.0 − 3.5 Hz.

Concerning the importance of vibration-induced muscle reaction in the activity of truck drivers, Müller-Limmroth (1961) has established that, when sitting, the muscular activity necessitates continuous reflex correction of the body posture because of the vehicle vibration. As time goes on, this correction activity can no longer be carried out precisely because of reflex fatigue. The weakening reflex action is caused by fatigue of the muscles and nerve pathways; however, he assumed that no true fatigue is involved, but instead an adaptation process (also see section "Vegetative and Biochemical Reactions," p. 60 ff).

In experiments carried out on the reclining person during vibration stress (Szameitat 1976), no changes were shown in the electric muscle activity of the m. trapezius in the neck-shoulder region, as opposed to resting. These results show that in the reclining person this muscle group can carry out no defensive reactions because of the lack of support possibilities.

Regarding the effect of continuing and interrupted vibration exposure, Scheibe (1979) was able to prove that during continuous stress, 60% of the time-varied changes in the electromyogram amplitudes were characterized as the fatigue process, as opposed to only 18% during interrupted stress. Frequent vibration interruptions, together with short exposure phases, therefore influence the muscles physiologically in the sense of reduced fatigue.

The breakdown of natural muscle reflexes as a result of vibration stress is reported in the section "Vegetative and Biochemical Reactions."

In summary, the previous knowledge on muscle reactions under vibration stress proves that each muscle group suitable for defense tries to counteract the vibration, with the goal of stabilizing the body or parts of the body as much as possible. If a part of the body finds itself in resonance vibration, it can only try to effect a change in spring constant and damping factor by means of changed muscle tension and, thus, also the vibration behavior of the body system involved. Whereas the musculature under resonance vibration shows especially high defense activity, under random vibration it reacts with increased static "pretension" in order to prepare to defend against the unpredictable vibration process. The possibility of carrying out such defense activity against mechanical vibration decreases as the posture changes from standing to sitting to lying down. Natural muscle reflexes, such as symptoms of disturbance in the vegetative nervous system, can be weakened by vibration or can break down.

5.3.2 Heart-Circulatory Function

Insofar as the effects of mechanical vibration on the heart-circulation system are concerned, a series of parameters such as heartbeat frequency, blood pressure, stroke volume, electrocardiogram, or plethysmogram would also have to show corresponding reactions. A review of the literature by Dressler (1969) and an overview by Seidel (1975) took this question into consideration. For methodological reasons, it is difficult to come to any conclusions based on the results, as the circulation parameters do not react specifically but can also be changed by stressors other than vibration.

During vibration exposure in the x-, y-, and z-directions, with increasing acceleration of 2.5 m/s² to 15 m/s², heart frequency is not significantly influenced, according to the research to Huang and Suggs (1965) and Schmitz (1959). Of course, Müller (1939) found a certain increase in heart frequency and a decrease in maximal blood pressure by means of measurements before and after vibration in a standing subject, but these experiments were conducted 44 years ago. When using a seated subject, however, he was not able to establish on average any significant changes in these two circulatory parameters. He points out the vast variation in results according to direction and magnitude.

In our own experiments using vibration with low-to-moderate intensity, such as that occurring in vehicles with good suspension, a small increase in heart frequency was established. However, this was reduced very rapidly soon after the beginning of the experiments. Since the subjects were accustomed to the test situation, this increase could not have been psychologically conditioned. Coermann (1965) observed a certain increase in heart frequency, but only at the beginning of vibration exposure of moderate intensity. He attributed this to subjective excitement. When reclining subjects were placed under vibration stress, however, Szameitat (1976) and Szameitat and Dupuis (1976) were not able to establish any significant increase in heart rate; there was, however, a decrease in the variation of heart frequency.

In the half-reclining position under stress from vertical sinusoidal vibration, Hood et al. (1966) found negligible heart-frequency changes at $a_{eff} = 4.2$ m/s²; at $a_{eff} = 8.4$ m/s² the changes were strong, increasing with exposure frequency from 4 to 10 Hz, to become smaller again above 10 Hz. Hood et al. (1965) were also able to show a rise in heart-rate frequency with increasing vibration acceleration in anesthetized dogs.

In long-term experiments on stress of simulated vertical tractor vibration in the sitting position of over 120-min duration, Christ (1973) found a gradual decrease in heart rate with time. This finding coincides with the results of Coermann (1965) that the heart rate decreases to values below the initial value after a prolonged period of exposure. Of course, a certain amount of habituation is also involved. Therefore, there is an analogy to the effect of acoustic stimuli in which, according to Jansen (1966), the heart rate is not changed in a sustained manner.

In practical driving experiments using tractor drivers, Bader (1967) found pulse-frequency increases only when extreme vibration stress occurred. Under simulated intense truck vibration ($a_{wz} = 2 - 4$ m/s²), Dupuis et al. (1974) showed in their experiments a short-term initial increase in heart rate of up to $20 - 30$ beats/min and then during the 45-min exposure, about 115% of the heart rate before exposure.

According to the interpretation of Coermann (1965), which is also in agreement with the opinion of Mandel (1963), a strong increase in pulse frequency only occurs in vibration stress near the endurance threshold in the frequency range below 20 Hz. This effect may have a psychological basis (in the sense of an anxiety reaction), or by augmented muscular activity.

Sjøflot and Suggs (1973) have shown that heart rate increases more strongly under rotational vibration than under linear vibration. This finding also corresponds with those of Dupuis et al. (1955) who, using nodding movement in the sitting position compared with vertical parallelogram movement, found an increase in energy consumption of up to threefold.

Regarding the data found under experimental conditions in the field, it must still be conceded that circulation depends for the most part on the special traffic conditions, individual capability, and on driving experience (Seidel 1975). The small increase in heart rate that was observed by Hoffmann et al. (1969a, b, 1976) in tractor drivers, in transporting people in ambulances, and in automobile drivers could therefore hardly reflect the effects of vibration stress.

The behavior of blood pressure when under vibration stress is apparently similar to that of heart rate. Loeb (1954) found no significant changes; Müller (1939) found a decrease in the systolic blood pressure. Coermann (1938, 1939) observed a blood pressure increase after strong vibration stress had been applied for $1 - 2$ h. Later, Coermann (1965) was of the opinion that the blood pressure amplitude (systolic-diastolic) increases even during relatively slight vibration stress, probably because of a rise in peripheral resistance. However, he could only determine direct vibration effects on the arterial blood pressure at $3 - 4$ Hz as a result of the relative movements of the diaphragm, but the changes were no greater than during forced breathing.

Schmitz and Boettcher (1960) decided to carry out experiments using several dogs because of the difficulty in measuring blood pressure directly and continuously, as this is only possible by using the invasive method of inserting catheters. Under sinusoidal stimulation between 1 and 8 Hz, with a constant displacement amplitude of 3-mm peak value with increasing frequency and therefore increasing acceleration, there was a rise in the systolic and a decrease in the diastolic blood pressure in the aorta at the level of the heart and right atrium.

On the basis of their human experiments (1966) and animal experiments (1965), Hood et al. believe that the vibration-caused cardiopulmonary changes − increase in blood pressure, heart frequency, oxygen consumption, and respiratory minute volume − which correspond to light muscular activity, can primarily be attributed to reflex muscle contraction.

There are only a few indications that changes may take place in the electrocardiogram (ECG), but these changes are not significant and cannot be correlated with specific physical vibration parameters (Dressler 1969). Experiments carried out in the seated position, under the influence of simulated truck vibration, have not indicated vibration-caused changes in the heart-action potentials, with chest-wall electrodes placed in accordance with the Wilson I, III, and V technique (Dupuis et al. 1974). ECG analysis of drivers of earth-moving machinery exposed to vibration were inconclusive with respect to vibration-caused ECG changes compared to a control group that was not exposed to vibration. No conclusion can be drawn with regard to these changes since they are nonspecific and could also have been caused by other stress factors (Köhne et al. 1982).

There are relatively few data on the influence of whole-body vibration on peripheral blood flow. In analogy to the hand-arm system exposed to vibration stress, vasoconstriction must be expected (Dupuis et al. 1982a), but the different vibration frequencies must also be taken into consideration. Müller (1939) carried out measurements of the skin temperature on the lower leg, which can be taken as a criterion for vessel contraction in the peripheral tissue. Even though it was demonstrated that vibration has no direct effect on these vessels, there was less peripheral blood flow in the extremities. Müller assumes from these results that well-known reflex procedures are responsible: they see that the abdominal organs and reservoirs fill up with blood under vibration stimulus and that the musculature also requires an increase in blood flow for its activity; thus less blood is available for the peripheral circulation. As found by Coermann et al. (1965), who measured the amplitude of the finger pulse at whole-body resonance vibration, an increase in peripheral resistance could also cause the rise in blood pressure.

According to Coermann (1965), the arterioles contract during noise as well as during whole-body vibration in most people, which results in a reduction of the peripheral blood flow. It is assumed that vasodilatation occurs in a few people, but otherwise, as a rule, there is strong vasoconstriction that varies according to interindividual differences. Immediately after the beginning of vibration stress, this vasoconstriction occurs and reaches its highest value after 1−5 min, after which it again decreases. Coermann's description does not

agree, however, with the information that has become available in the meantime regarding hand-arm vibration: vasoconstriction is longer and more continuous, with simultaneously exerted grip strength (Dupuis 1982).

In summary, it must be stated that so far neither the causes nor the mechanisms are known with respect to the reaction of the cardiovascular system to
whole-body vibration. The increases in heart frequency are apparent only under very high vibration stress, the changes in blood pressure are inconsistent,
and the ECG changes cannot be significantly differentiated. There has been
little research on the disturbances in peripheral blood flow due to vasoconstriction, but these disturbances are apparently weaker and last for a shorter
period of time than in hand-arm vibration, for which the blood-flow disturbances are marked due to the exertion of the grip forces in the fingers.

5.3.3 Respiratory Function

The influence on muscular activity, which is dealt with in the section "Muscle
Activity" (see p. 51), can also be explained by the increase in energy transfor-
mation. Furthermore, during vibration stress an increase in ventilation can also
be expected due to the vibration behavior of the organs in the abdomen and
thorax. Oxygen consumption, respiratory minute volume, and respiratory frequency are the natural physiological parameters to provide these data.

In earlier experiments (Dupuis et al. 1955), it had been established that the
energy consumption (kJ/min) of the subjects was strongly dependent on vibration, which was a function of the various kinds of drivers' seats. The caloric expenditure was measured by means of respiration procedures described by Müller and Franz (1952) and by using the gas analysis technique according to
Douglas-Haldane. In experiments using five subjects and one repetition of each
test, the caloric consumption was increased up to 330% during the use of a seat
with poor suspension and strong nodding vibration in comparison to a paral-
lelogram-guided driver's seat with good suspension and shock-absorption
properties.

In other experiments, respiratory behavior under the influence of sinusoidal
vibration was investigated (Dupuis 1969). Nine subjects were exposed to sinusoidal vibration at a frequency range of 1.1 – 10 Hz, with frequency steps of
1 Hz above 2 Hz. The displacement amplitudes were chosen so that in all experiments there would be a constant "perception magnitude" K-value of 25, in
accordance with the early VDI recommendation 2057 (1963). The purpose was
to check whether the frequency-dependent, approximately acceleration-proportional course of the K curves of up to 10 Hz also corresponds to objective
physiological criteria. These K curves are essentially based on the experiments
of Dieckmann (1957), according to Bobbert (1967). The biodynamic behavior
of the body parts (section "Biodynamic Reaction on Vibration," p. 14) has already been shown to deviate a great deal from the early VDI recommendation
of 1963.

Using a respirometer with a mouthpiece with a valve and by closing the
nostrils with a clamp, the amount of air exhaled was measured during 5-min

vibration. Simultaneously, the breathing rate was measured by means of an inductive pressure indicator in line with the respirometer and recorded. Thus, the
respiratory frequency was obtained for each minute of the experiment.

All subjects had to grade their subjectively felt perception on an established
scale at the end of the experiment. In spite of the well-known problems with
such subjectively scaled classifications, an evaluation immediately after the experiment appeared to be possible in comparison to the previous test, and
should result in a correlation between respiratory behavior and subjective perception.

The results from this series of experiments are presented in Fig. 31. At all
vibration frequencies, with the exception of 1.1 Hz, the respiration rate curve is
approximately 10% – 20% below the resting value. These results confirm those
of Müller (1939) who, in his series of experiments, had established in general
that the respiratory rate greatly decreases. However, Gaeumann et al. (1962)
have found decreases in respiration frequency as well as increases.

As shown in Fig. 31, the respiratory minute volume generally follows a
course that contrasts with the respiratory frequency. It climbs from 2 Hz to 5 Hz
and then again shows a falling tendency. Corresponding to the constant K value
of 25, in accordance with VDI 2057 (1963), the acceleration was approximately
constant up to 5 Hz and increased only slightly above 5 Hz. This means that the
respiratory minute volume under low frequency vibration may be considered as
"vibration-induced hyperventilation."

The frequency-dependent course of respiratory minute volume shows a tendency that is very similar to the course in subjective perception (not presented
here). The correlation coefficient is high, with $r = 0.871$. From these results, it
is quite obvious that in the frequency range of < 10 Hz, the frequency-dependent course of "perceived magnitude K," in accordance with VDI 2057

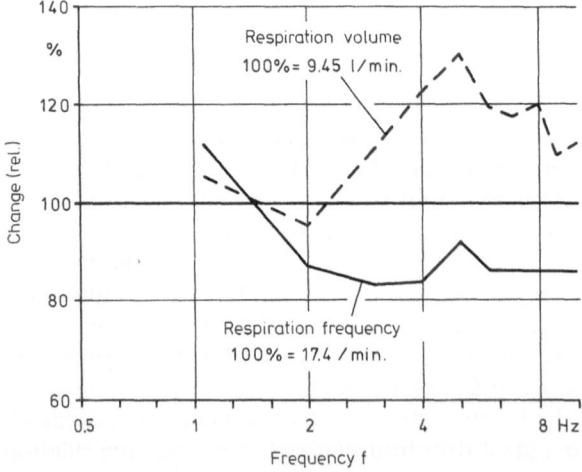

Fig. 31. Frequency-dependent relative changes in the respiratory minute volume and the
breathing frequency, with reference to the resting value, in the seated position — rms-
weighted acceleration $a_w = 1.25$ m/s² constant (Dupuis 1969)

(1963), in no way correlates with respiratory minute volume and subjective perception. These results have also contributed to the revision of VDI 2057 (1981) regarding a change in frequency-dependent evaluation.

Gaeumann et al. (1962) established 6 Hz as the maximum in induced hyperventilation but, in contrast, found that vibration has a calming effect on respiration at 2 Hz. In general, Müller's experiments (1939) had similar results: when frequency and acceleration (since the displacement was kept constant) were increased, there were increases in respiratory minute volume. Huang and Suggs (1965) also found a strong increase in ventilation with increasing acceleration. According to Lamb et al. (1966), the extent of hyperventilation varies individually.

Oxygen consumption also shows clear changes under vibration stress. This point is supported by a series of experiments (Müller 1939; Coermann 1938, 1939; Duffner et al. 1962; Gaeumann et al. 1962; Hood et al. 1966; Berthoz 1969). A vibration-caused increase in oxygen consumption was reported in all reports. Duffner et al. (1962) had already established significant increases like these at about 1 m/s² acceleration, which increased further with additional acceleration. Acceleration-proportional increases in oxygen consumption were also found by Huang and Suggs (1965) in seated subjects exposed to vibration stress in the vertical direction but not, however, in the horizontal directions. The frequency range in which oxygen consumption increases most clearly has been reported in several studies to be between 2 and 6 Hz. These results can certainly be explained by the increase in muscular activity (see section "Muscle Activity," p. 51).

In summary, it has been shown that hyperventilation can occur as a result of vibration; this is especially apparent in the resonance range of the human body and when there is great acceleration. Since respiration, however, is more directly excited by the influence of vibration than energy consumption, hyperventilation is caused by energy expenditure only to a small degree. This has been shown by Müller (1939) in considering the quotient

$$\frac{\text{Ventilation}}{\text{O}_2 \text{ Consumption}}$$

as a function of frequency. The main cause for hyperventilation is the passive movement of the diaphragm and abdominal wall, which is driven by vibration of the viscera. It leads to a kind of artificial respiration. Apparently this is mostly due to ventilation of the resting volume, as already pointed out in the discussion on the vibration behavior of the internal organs (section "Vibration Behavior of Internal Organs," p. 39).

Under vibration stress, blood gas analysis showed a slight (nonsignificant) increase in oxygen partial pressure, pO_2, whereas no changes in carbon dioxide pressure, pCO_2, and pH values could be established (Dupuis et al. 1974).

5.3.4 Vegetative and Biochemical Reactions

Indirect effects on the vegetative nervous system can also be expected as a result of mechanical vibration. In the discussion on circulatory behavior (section "Heart-Circulatory Function," p. 54), it has already been pointed out that the vasomotoric system may be affected. In addition, in the discussion on muscular activity (section "Muscle Activity," p. 51), the inhibition of natural muscle reflexes has also been pointed out. Both of these functions are controlled by the vegetative nervous system. Furthermore, potential effects on the endocrine system should be explored.

When using subjects exposed to vibration above 20 Hz, Coermann (1938, 1939) observed that after a short time there were no patellar reflexes and that these reflexes returned sometime after the end of exposure. Goldman (1948a) and Goldman and von Gierke (1960) also pointed out that normal reflex action may be blocked as a result of vibration. Loeckle (1941) was able to show in an animal experiment that this loss is caused by stimulation of the receptors of the smooth musculature, which results in blockage of the reflex arch and synapse. However, the normal voluntary movement of the muscles involved remains uninfluenced. In seated subjects exposed to vibration at 18 Hz and 2.5 m/s² acceleration amplitude, Roll et al. (1980) have shown a decrease in the tendon reflexes in the region of the lower extremities. According to Seidel (1975), loss of reflexes or weak reflexes are particularly apparent at a vibration stress of > 30 Hz, as has been well established in various other experiments (Borscevskij et al. 1963; Coermann 1938, 1939, 1962; de Gail et al. 1966; Goldman 1948a; Guinard 1960; Loeckle 1941). In contrast, however, Guinard and Travers (1959) were not able to detect any reflex loss as a result of intensive whole-body vibration at 2–10 Hz.

The biceps and Achilles tendon reflexes are also said to be reduced during vibration, but probably with less sensitivity than the patellar reflex.

The degree and speed of the impairment of these reflexes are apparently dependent on the frequency, amplitude, and duration of the vibration. Guignard and Travers (1959) believe that the effects are controlled by the higher nerve centers. On the whole, however, the current knowledge on the type, extent, and probable causes of vibration-dependent decreases in reflex responses is regarded as insufficient.

As pointed out in the section "Heart-Circulatory Function" (p. 54), during vibration stress the reflex processes can effect a change in blood distribution in the sense that there is a decrease in the blood flow to the peripheral vessels in order to provide blood to certain organs and the musculature. This is based on an assumption by Müller (1939). According to Selye (1949), these processes must be regarded as one of the components of an alarm reaction: an atavistic reaction that is actually inadequate during vibration stress since the blood requirement of the musculature is relatively low compared to the energy consumption for continuous effort.

Therefore, the vegetative nervous system must also be held responsible for the decrease in blood flow to the skin. This has already been shown to be an effect in the case of acoustic stimulation – noise (Jansen 1966). In our own measurements of thermal conductivity and skin temperature, as well as the finger pulse, a significant decrease in the peripheral blood flow was established as a result of noise and of hand-arm vibration (Dupuis 1982a).

The electrical characteristic of the skin is connected with its blood flow and penetration with tissue fluids. Thus, in our own experiments 28 years ago, in which we evaluated the vibration effects of tractor seats, we showed that the electrical capacity of the skin is much lower when seats have poor suspension (i.e., stronger vibration stress) than when there is good suspension (Dupuis et al. 1955). There was also a certain correlation between the extent of the electrical skin capacity and a person's subjective perception of the vibration. Later, these findings were confirmed experimentally by Dieckmann (1957), who based these findings on, among others, the evaluation scale (K value) that he developed. With the help of finger plethysmography, Yonekawa (1978) was able to show an increasing reduction of peripheral blood flow during increasing vibration stress. No further unambiguous data are known regarding the vibration-dependent changes of the vasomotoric system under whole-body vibration.

It is often asked whether mechanical vibration in general acts as a stimulus to the secretion of the gastric juices in a way similar to ship movement (Bödecker 1937). At first, Dupuis and Christ (1966) pursued this question from the physiological nutrition point of view: is the stomach emptied more rapidly if vibration is applied? To clarify this question experiments were carried out in which X-ray films were taken of the stomach after standardized food intake at 30-min intervals. However, no significant data were obtained. In addition, experiments carried out by Suzuki et al. (1959) on the same question also provided no definitive results. If the experiments show any tendency at all, then it appears that mechanical vibration acts as a stimulus to increase the formation of gastric juices.

In further studies using six subjects, telemetric determination of the pH values of the stomach content using the Heidelberg capsule, according to Hochberg et al. (1964), was used to find the potential change in acidity of the gastric juices or acid-formation capacity (Nöller and Khodabakhsh 1964) during exposure to vibration (Dupuis 1969).

Not all experiments could be evaluated, as in one subject the Heidelberg capsule had already gone into the duodenum after 10 min. When compared with the experiment at rest, there was more or less rapid transfer to the duodenum during vibration stress in some subjects; this transfer is recognizable when the pH value suddenly climbs to about 5–6. During the influence of vibration, other subjects showed a strong fluctuation in pH value and rapidly reached a high level of acidity. In general, then, it was not possible to confirm the above-mentioned hypothesis that secretion increases as a result of vibration since the experimental results were too variable.

Further indications of the influence of vibration on the digestive processes are given in the works by Usutani et al. (1965), who have carried out animal experiments on rats under the effect of vibration (7-Hz frequency, 10-mm displacement amplitude). There was a decrease in body weight in the animal group subjected to vibration in contrast to the control group, and the amount of food eaten per body weight unit was also significantly less in the animals under the influence of vibration.

Although it must also be assumed that mechanical vibration influences the endocrine system and blood count and that biochemical changes are evoked, there are no convincing results to support these assumptions. With regard to this question, Coermann (1965) cites a few studies that are mostly American (Blivaiss and Foa 1964; Cope and Polis 1959; Mandel et al. 1962). In rhesus monkeys under the influence of extremely high vibration stress, changes in serum-glutamic acid-oxaloacetic acid-transaminasis (SGOT) were found that should represent a secondary effect of nonspecific stress. No SGOT changes were found in human subjects, however, although both groups were exposed to vibration for 20−60 s at the subject tolerance level in the range of 4−9 Hz.

Ivanovitsch et al. (1981) were not able to establish any significant changes in succinate-dehydrogenase (SucDH), lactate-dehydrogenase (LDH), and adenosine-triphosphate (ATP) during animal experiments on rats subjected to vibration stress (10 days, 2 h daily, 0.1-mm amplitude, 100 Hz).

Since Tippelmann (1964) had determined that in various kinds of stress the number of eosinophile blood cells per cubic millimeter of blood is reduced, Coermann et al. (1965) used this research procedure on human beings subjected to the effects of vibration. Samples were taken and counted both before and 15-20 min after the vibration stress. However, no clear correlation could be established with regard to the type of vibration stress.

In experiments by Dupuis et al. (1974), the blood corpuscles were practically unchanged during vibration stress, with the exception of a slight median increase in the leukocytes. However, this was not confirmed statistically.

Numerous laboratory parameters have been measured in drivers of earth-moving machinery before and after exposure to vibration, as well as in a control group (blood sedimentation rate, albumin, globulin, enzyme, cholesterol, triglyceride, creatinine, uric acid, potassium, calcium, blood sugar). The results showed no differences between the groups that could be statistically confirmed; instead, the values were in the range of the physiological norms (Köhne et al. 1982).

Puschkina (1961) observed a tendency toward hypoglycemia, a drop in the albumin-globulin coefficient, a reduction in the amount of ascorbic acid, and an increase in the globulin fractions. However, no particularly striking changes in rest nitrogen, bilirubin, creatinine, or calcium were found.

With regard to the question of the effects of emotional stress, no significant differences were found in the distribution of catecholamine in subjects exposed to vibration compared to control experiments (Starlinger et al. 1969). This is supported by subjective findings, which indicate that vibration stress causes no significant emotional excitation (Hawel 1969).

Moreover, there are indications that as a result of direct exposure, mechanical vibration can evoke an increase in bone growth. In his animal experiments on the "shaking theory" in young rabbits, Sergel (1983) found noticeable differences between measurements on the right and left side of the lower jaw, as well as local bone increase in the area of the lower jaw directly exposed to vibration. However, the results are not sufficient for use as proof of the shaking theory (see section "Biological Prevention and Control Mechanisms Against Mechanical Vibration," p. 12).

It is well known that certain sensory activities, excitation states, sleep behavior, and certain illnesses may lead to changes in the electrical curve picture of the brain, which can be proved by means of electroencephalography (EEG). Thus, attempts have also been made to show EEG changes during vibration stress (Adey et al. 1963; Borscevskij et al. 1963; Kozlov and Kiseleva 1971; Nickolson and Guignard 1966). In this regard, however, no causes have been discovered for the EEG changes during vibration in animal experiments or in human experiments. It was not possible to show consistent trends for these changes, because the range of EEG reactions is too wide due to individual differences.

According to Seidel (1975), only a few Soviet and Bulgarian publications indicate that EEG changes occur as a result of long-term whole-body vibration on the job (Spilberg 1962; Petrov 1968; Ruppe 1971). So far it has not been confirmed that EEG changes occur as a result of vibration stress. However, Yamazaki (1977) used EEG to determine the depth of sleep during the influence of vibration. He discovered that sleep is not influenced by vibration stress of < 0.01 m/s².

In summary, it has been established that there are indications that the natural muscle reflexes controlled by the peripheral nervous system and the peripheral blood circulation can be reduced by the stress of whole-body vibration. In general, vibration-caused changes in the electroencephalogram, the blood count, and the endocrine system, or changes of a biochemical nature have not been generally demonstrated. Instead, the changes observed appear to lie in the normal physiological range.

5.3.5 Sensory Functions

Mechanical vibration can be expected to have an influence on the sensory organs and the psychomotor system. In this connection, the direct mechanical effects on these organs and the indirect neurophysiological effects must, however, be differentiated, although these effects often cannot be distinguished by the measurements applied.

With regard to direct effects, it is well known that vibration can impair speaking ability, which is caused by resonance vibration in the trachea and bronchial tubes (Magid and Coermann 1960; also see Fig. 29). Nixon (1962) has further established, with regard to this question, that pitch may be raised and speech duration extended by about 15%, dependent on the vibration frequency. When speaking ability is reduced, intelligibility can be impaired, which can have practical consequences in telecommunications.

From the point of view of direct influences, it must also be asked whether or not mechanical vibration (which belongs, according to its frequency, to the transition zone between infrasonic and sonic waves) has an influence on hearing. Airborne sound of high intensity, such as continuous noise of about 150 dB near the exhaust pipe of jet power plants, the short-term explosions when munitions and rockets are fired, and supersonic booms, all contain in their frequency spectrum high energies in the infrasonic range. Because of the difference in acoustic impedance of the body surface and its underlying tissues, this energy generally does not penetrate deeply in the body but is to a great extent reflected. However, the high-intensity, low-frequency components of the air waves can produce mechanical vibration in the head, which is transmitted directly to the inner ear by means of bone conduction, as well as to the other sense organs.

5.3.5.1 Acoustic Perception

There is still no clear evidence as to whether low-frequency vibration can produce lasting impairment of hearing or permanent threshold shift (PTS) by means of bone conduction. Animal experiments carried out by Nakamura (1941) seem to have proved this, however. On the other hand, it must be taken into consideration that, in practice, such vibration is always coupled with high-frequency components of the audible range.

According to the review by Dupuis (1979), the question of whether mechanical vibration has any kind of deafening effect on hearing has not been consistently answered. According to this review, Teare and Snyder (1963) were not able to provide any proof that whole-body vibration alone changes the auditory threshold. On the other hand, Okada et al. (1972) found that mechanical vibration alone (5 Hz 5 m/s²) can result in significant temporary threshold shift (TTS) at a testing frequency of 1 and 5 kHz.

For this reason, the following question is important: Is the decrease in hearing from a combination of vibration and noise stress more pronounced than from exposure to noise alone? On the basis of his experiments (disregarding a few special effects), Sommer (1973) observed that the TTS was not significantly influenced by the combination of noise and vibration.

In contrast, the experiments by Yokohama et al. (1974) have shown that vibration and noise together cause a higher degree of TTS at 4 kHz (TTS: 12 dB) than noise alone (TTS: 5 dB) and that recovery takes longer.

Using five experimental subjects, Okada et al. (1972) conducted research on the combined effects of whole-body vibration and noise in the sitting position (101-dB broad-band noise, 5-Hz vibration with 5.0 m/s²). It was found that the auditory threshold was shifted approximately 5 dB at 1 and 4 kHz, as opposed to the effects of noise alone.

The results by Okada et al. caused Pfander (1978) to conduct control experiments, in which 104 subjects were subjected to the noise of a chain-driven vehicle [99−102 dB (A)] for 30 min. For comparison, the subjects in addition were exposed to vertical vibration (5 Hz, 5.0 m/s²). When compared with noise exposure alone, the additional vibration exposure showed no stronger effect on $TTS_{2 min}$ and on the recovery time.

The auditory threshold shift in tractor drivers has been studied by Pinter (1975), who used as control subjects workers in the furniture industry. Both groups were exposed to equivalent noise levels of 90−98 dB (A). A stronger reduction in hearing was shown in the tractor drivers. The $TTS_{2\,min}$ values measured were above those calculated, and the TTS values lay above the expected values. Pinter concluded that the effect of vibration intensifies the effects of simultaneous exposure.

To summarize, one can conclude from the different results of the data presented that it is so far not possible to draw generally valid, transferable conclusions as to whether mechanical vibration can cause deafness or hearing difficulties. This is due to the fact that the stress levels chosen by the authors varied a great deal with regard to sound intensity, spectrum, acceleration frequency, and exposure time.

5.3.5.2 Equilibrium Regulation

When driving motor vehicles or flying airplanes and helicopters, any influence of mechanical vibration on the vestibular apparatus is very important. Various experiments have been carried out on this question (Loeckle 1950; Lebedeva and Puskin 1950; Simons and Schmitz 1958; Schmitz and Simon 1959; Coermann et al. 1962). Lindner (1962) summarized the knowledge available at that time on vibration-caused equilibrium disturbances.

Coermann et al. (1962) and Schmitz and Simons (1959) have determined the ability of the upper body to recognize and compensate for deviations from the vertical position, with the eyes closed or covered, after exposure to vertical vibration. Schmitz and Simon, who carried out these tests before and after the influence of vibration, found no significant results at vibration frequencies of 2.5−3.5 Hz and acceleration between 1.5 and 3.5 m/s². However, Martin et al. (1980) have been able to show a significant increase in the amplitude of swaying movements in the standing position after exposure to vibration stress when standing and sitting with 5.0 m/s² amplitude at 18 Hz.

Coermann et al. tested subjects during vibration stress and established that they were under especially strong physical stress at the resonance frequencies of the whole body, which distracted them from the test task presented. A great effort of will, however, enabled a few subjects to compensate for this stress, so that vestibular efficiency was not influenced. In the opinion of Coermann (1965), there is apparently no direct influence on the equilibrium organ during translational vertical vibration in the frequency range of 1−20 Hz.

Nevertheless, these results differ from the opinions of a few Russian authors (Borscevskij et al. 1963; Lebedeva and Puskin 1950; Menshov 1962, 1967), as pointed out by Seidel (1975). In addition, O'Hanlon and McCauley (1974) have been able to evoke strong vestibular responses during pure vertical vibration, which finally led to kinetosis.

Guignard (1965) believes that disturbances in the regulation of body posture, as measured by stabilography, are probably related to influences on the vestibular and spinal mechanisms as a result of vibration stress.

5.3.5.3 Kinetosis

In our own experimental field trips with chain-driven vehicles producing strong nodding stress and simultaneously demanding psychomotor performance, physiological overload with nausea and disturbance of consciousness was observed. Therefore, the following question is apparent: Under which physical vibration parameters can kinetosis be evoked?

In the Federal Republic of Germany, the general term "kinetosis" (motion sickness) is used to include individual concepts, such as air sickness, car, sea, railroad, elevator, and merry-go-round sickness, which are caused mostly by very low-frequency vibration with great displacement amplitude, but are also influenced by additional factors. These illnesses are of an acute nature and do not always begin immediately after the onset of stress, but often after a longer period of exposure (minutes to hours). Whereas until the end of the last century only sea sickness was referred to as kinetosis (for the history of sea sickness, see Schadewaldt 1967), it has grown in general importance as the result of the rapid development of transportation.

There is a multitude of literature on kinetosis, generally in connection with a discussion on the pathogenesis, symptoms, prophylaxis, and therapy, and less with regard to its physical vibration parameters (Geller 1940; Rudat 1952; Glaser 1953; Johnson 1954; Goethe 1954, 1956, 1967, 1973; Goethe and Fischer 1957; Reinhardt 1959; Spangenberg 1962; Dowd 1965; Yules 1967; Money 1970; Goto and Kanda 1977). Despite the comprehensive literature summarized only briefly here, the state of knowledge regarding the origin of the illness and the effect of various stress factors on the different receptor systems is still not complete.

Goethe (1973) points out that of the receptor systems, the vestibular apparatus is certainly the most important organ regarding orientation in space but that for the perception of acceleration stimuli and the triggering of kinetosis, almost the entire somatosensory system is involved. In addition, the mechanoreceptors of the skin, of the connective tissue, and the joints play a role, as does the musculature of the skeletal system. It is probable that the afferent impulses from the region of the suspension of the abdominal organs at the mesenterium also have a certain importance in signaling changes in the abdominal pressure. Also, optical information alone is apparently of considerable importance and can lead to optical disturbances in coordination, e.g., in stereo wide-angle film displays with strong motion events or in connection with low-frequency mechanical vibration. Finally, the olfactory system appears to play a decisive role, for example, in sea kinetosis, because certain odor stimuli (oil fumes, exhaust, food smells, etc.), which would under normal conditions be perceived as insignificant, can have a strong triggering effect at sea, resulting in kinetosis with vomiting.

So far, the kind of mechanical vibration that leads to kinetosis has not been clearly defined. Goethe (1973) is of the opinion that vertical acceleration, frequency, and motion pattern trigger sea kinetosis and that rotational vibration, on the other hand, plays a lesser role. In contrast to this, the assumption is well founded that rotational vibration in the low-frequency range leads to the stimulation of the vestibular apparatus. Stimulation of the semicircular canals is thus also probably partly responsible for the development of kinetosis (Bödecker 1937; Geller 1940; Dieckmann 1961). Since this effect appears to be greatly dependent on the individual, however, still other triggering factors must be operative, and it does not necessarily in all cases have to result in kinetosis.

There have been interesting results based on observation of the incidence of kinetosis in cadets who were subjected to sea-going ship vibration under various sea states for different time periods. Goto and Kanda (1977) found that the risk of becoming ill increased as the vibration frequency decreased from 0.5 to 0.1 Hz, with equal vertical acceleration. These data also correspond to the frequency-dependent curves by Yonekawa and Miwa (1972) for equal subjective perception. (Also see section "Subjective Perception," p. 77.) Therefore, this means that kinetosis can be expected to appear only at vibration frequencies below 0.5 Hz (primarily below 0.3 Hz) and that the effects are intensified with decreasing frequency at a constant acceleration. (Although the vibration on ships mostly consists of roll and pitch, for instrumentation reasons only the vertical vibration components were used as measures of stress.) The frequency-dependent international evaluation curves (ISO 2631, Addendum 2, 1982) are therefore in line with this trend.

As shown in Table 7, Goethe (1973) subdivided the symptomatology of the illness into five stages according to degree of severity; Kanda et al. (1977) used three stages.

In their experiments on cadets, Kanda et al. (1977) established a decreasing incidence with increasing degree of illness. Under approximately constant vibration acceleration, the maximum incidence occurred after exposure of 2−3 h. The relationships between acceleration and incidence of kinetosis are shown in Fig. 32 for the three stages of illness severity. The curve for grade III by Kanda et al. (1977) is comparable to the results of the laboratory experiments by O'Hanlon and McCauley (1974). The differences resulted from the fact that in

Table 7. Grade of kinetosis symptoms

Symptom	Grade according to	
	Kanda et al. (1977)	Goethe (1973)
None	0	0
Feeling somewhat unpleasant, fatigue, exhaustion, lack of appetite, pale, little desire to work	I	1
In addition: nausea, very weak, feeling unwell, dizziness, absolute inability to perform any activity	II	2
Very nauseated, feeling unwell, vomiting	II	3
Heavy vomiting, constant urge to urinate and defecate, feeling very ill	III	4
Continuous vomiting, feeling like death, pronounced adynamia, lack of ability to take part in anything	III	5

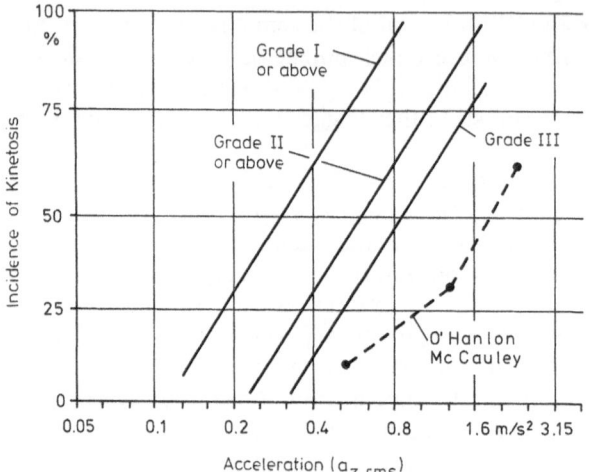

Fig. 32. Maximal incidence of kinetosis according to various grades of severity (see Table 7) and related to vertical rms acceleration a_z (Kanda et al. 1977; O'Hanlon and McCauley 1974)

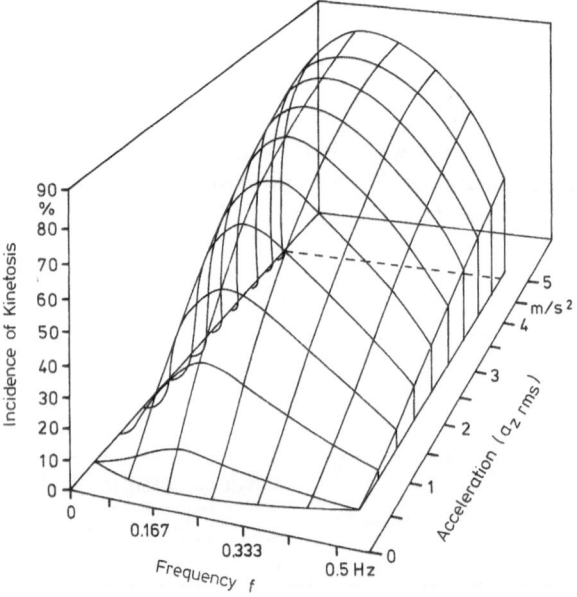

Fig. 33. Correlation of the incidence of kinetosis with frequency f and the rms acceleration a_z in the vertical direction (O'Hanlon and McCauley 1974)

the laboratory experiments, sinusoidal vertical vibration was used for exposure, whereas the cadets on the ships were also exposed to rotational vibration, which was not measured.

O'Hanlon and McCauley (1974) have established the empirically derived relationships of the incidence of kinetosis to acceleration and frequency of vertical sinusoidal vibration (Fig. 33). It is apparent that the incidence increases with decreasing frequency down to 0.17 Hz and with climbing acceleration.

There is little clarification of the problem of adaptation to motion stress, which should probably be regarded as habituation rather than as adaptation, according to Goethe (1973). Prophylactic measures have to do with a reduction in vibration and stabilization, the conduct of the person affected, and pharmacological treatment.

5.3.5.4 Visual Perception

It has already been pointed out that the presentation of visual information in connection with low-frequency vibration can result in disturbances in coordination. Thus, when a high degree of visual proficiency is demanded, the question of whether mechanical vibration can negatively influence this visual capability or not plays a great role. It is understandable that this problem is of special importance in aviation and space medicine, and most of the experiments in this area have been conducted in the United States. Griffin and Lewis (1978) and Dupuis and Hartung (1980a) have carried out an analysis of the literature that encompasses about 70 original studies. Here only the most important experiments or review articles should be mentioned (Coermann 1938, 1939; Simons and Schmitz 1958; Drazin and Guignard 1959; Guignard and Irving 1960; Teare and Parks 1963; Dennis 1963, 1965; Oshima 1963; Snyder 1965; Taub 1966; Rubinstein and Kaplan 1968; Grether 1971; Lee and King 1971; Shoenberger 1974; Schmidtke 1974, 1975; Griffin 1975b, 1976a; Ohlbaum 1976; Barnes 1979; Lewis and Griffin 1980; Dupuis and Hartung 1980a; Dupuis 1981d; Hartung 1983).

The presentation of generally valid data with regard to this question is rendered even more difficult, as numerous technical stress factors can be operative. There are three different working conditions in which the taking up of visual information and its processing can be hindered by mechanical vibration:

Person stationary – Sight object vibrating
Person vibrating – Sight object stationary
Person vibrating – Sight object vibrating

It is to be expected that visual perception is impaired in the first combination and that the impairment increases in the second for equal-strength vibration stress. This is particularly valid for vibration frequencies which result in whole-body or partial-body resonance in man. In the third combination, the difficulty of visual perception is determined, in addition, by the phase lag between the person and the sight object (Dupuis and Hartung 1980a).

Furthermore, a distinction must be made between translational or rotational vibration and the direction with respect to the human body. Because of the different biodynamic transmission paths to the eye, the site of vibration induction into the body (e.g., extremities, seat, head) and body posture (standing, sitting, or lying down) are important. The preciseness of visual perception is determined by the frequency-dependent relative vibration movement between the eye and site object, whereby physiologically possible compensatory secondary movements of the eye are possible but limited.

Many experimental results cannot be compared if the vibration is only measured at the induction site into the body but not in the vicinity of the eyes. Measurements of the vibration amplitude near the eye are required if data on the impairment of visual capability are to be related to the physical stress.

In this regard, the influence on visual acuity and the visual recognition time during mechanical vibration have been investigated (Dupuis and Hartung 1980a; Dupuis 1981; Hartung 1983). By means of separate experimental series, the influence on visual perception during vibration stress in the x-, y-, and z-directions, with a constant acceleration of $a_{eff} = 1.0$ m/s² at the head, was tested on ten subjects. The vibration stress was produced in the z-direction over the frequency range 0.8 – 12.5 Hz by means of an electrohydraulic vibration simulator with vibration induction to the seat of sitting subjects. For the x-, y-, and z-directions and the frequency range 12.5 – 80 Hz, the stress was induced by means of an electromagnetic vibration simulator with vibration induction at the cranium.

To achieve reproducibility of the experimental procedure, the Landolt-ring optotypes served to determine visual acuity in accordance with German standards (DIN 58220, p. 1, 1974). These optotypes were presented to the subject in blocks of eight each, with a defined degree of difficulty at a distance of 5.0 m. Seven lines of projected Landolt rings always corresponded to visual acuity values between 0.8 and 1.6.

Tachistoscopic tests were done with Landolt rings, as well as with four-digit groups of numbers as optotypes (as used in clinical practice). Absolute comparison of both tests was not possible, since Landolt ring data (Schober 1976) refer to the minimum separable, but numbers to the minimum recognizable.

Only slight, nonsignificant reductions in visual acuity were shown under vibration stress at low-frequency between 5 and 8 Hz (Fig. 34). Using considerably higher vibration stress corresponding to $a_{wz} = 3.5 - 5.0$ m/s² (K values of

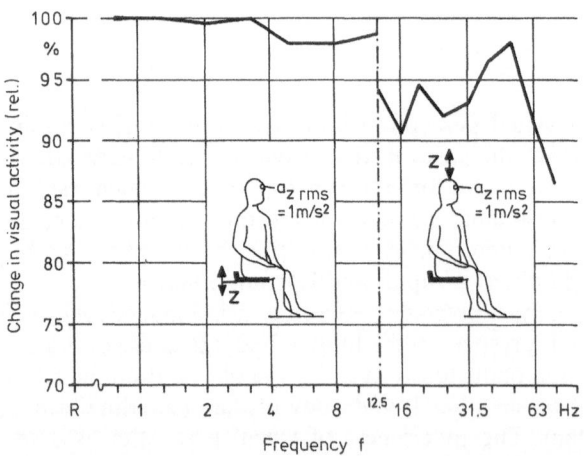

Fig. 34. Relative changes in visual acuity related to the resting value, under vertical vibration of the head and with rms acceleration $a_z = 1.0$ m/s². 0.8 – 12.5 Hz: vibration transmission to the buttock (buttocks); 12.5 – 80 Hz: vibration transmission to the head (Dupuis and Hartung 1980a)

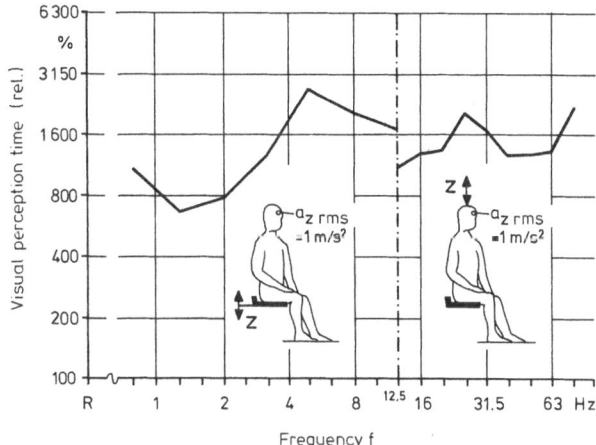

Fig. 35. Relative changes in visual perception times for number combinations, related to the resting value, under vertical vibration of the head with rms acceleration $a_z = 1.0$ m/s². 0.8–12.5 Hz: vibration transmission to the buttocks; 12.5–80 Hz: vibration transmission to the head (Dupuis and Hartung 1980a)

between 70 and 100), Schmidtke (1974) established a reduction in the resolving power of about 30% in the frequency range 1.5–4.5 Hz. In contrast to rest conditions, clear significant decreases in visual acuity are found at higher frequencies, which are especially pronounced between 16 and 31 Hz, as well as at 80 Hz. These data correspond to the experience of crew members of certain types of helicopters, in which 27-Hz vibration led to considerable difficulties in reading maps (Hartung 1983).

The results in tachistoscopic tests showed similar frequency-dependent tendencies, which demonstrated in general, during vibration stress, strongly increased visual perception times, as opposed to rest conditions (Fig. 35). These perception times increased, for example, 6 to 20 times for the four-digit number groups. In addition, different vibration frequncies showed an especially high increase in visual perception time, at 5–8 Hz and around 25 Hz.

Whereas very slight changes in visual perception were noted in the horizontal x-direction (dorsoventral), these changes proved to be especially high in the horizontal y-direction (shoulder-shoulder), in the middle frequency range 25–40 Hz. Visual acuity was affected (Fig. 36) and decreased by up to 20%, and the visual perception times (Fig. 37) increased up to 30-fold.

The experiments of Lewis and Griffin (1979a) have further shown that with increasing acceleration of the vibration stress, the number of errors in reading of numbers greatly increases.

The influence of vibration stress on capability may have various causes in the biodynamic, physiological, and/or psychological realms, and they may partly overlap. To these causes belong:

Insufficient compensatory secondary movement of the eyes at low frequencies

In translational vibration, compensatory eye movements for the purpose of target pursuit are only fully possible up to approximately 2.0 Hz; in addition,

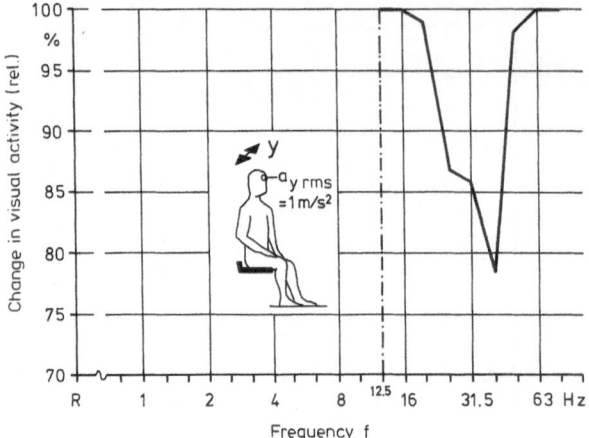

Fig. 36. Relative changes in visual perception, related to the resting value, under horizontal vibration of the head with rms acceleration $a_y = 1.0$ m/s² at frequencies from 12.5 to 80 Hz (Dupuis and Hartung 1980 a)

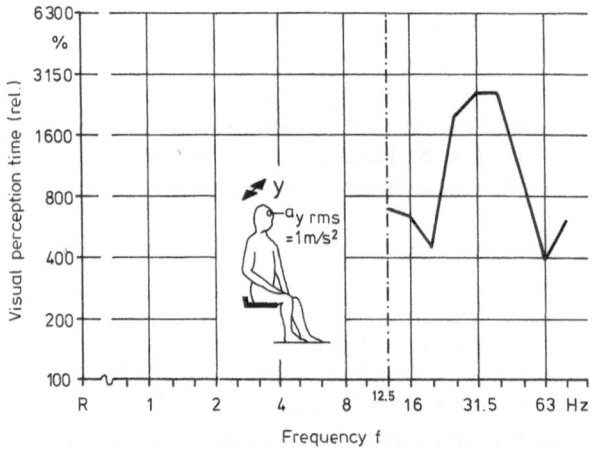

Fig. 37. Relative changes in visual perception times for number combinations, related to the resting value, under horizontal vibration of the head with rms acceleration $a_y = 1.0$ m/s² at frequencies of 12.5–80 Hz (Dupuis and Hartung 1980 a)

there is a displacement-dependent factor (Drazin and Guignard 1959). Above 4.0–5.0 Hz, there is a breakdown of the compensatory action. It is not clear whether this capacity is limited more when the person is vibrated or when the object is vibrated (Guignard 1965).

Vibration transmission through the trunk and head, including rotation and resonance of the head

Drazin and Guignard (1959) have observed difficulties in visual perception caused by whole-body resonance. The aforementioned results of Dupuis and Hartung (1980a) also suggest that there is a reduction in visual performance during resonance of the trunk, but that this reduction is not significant. During low-frequency rotational vibration with a vibrating object, Benson (1972) found a strong reduction in visual performance of the stationary human subject. However, when the subject was under vibration stress and the sight object stationary, this reduction was appreciably less. Benson considered these results to stem from vestibulo-ocular reflexes, whereby the eye movements during rotational movements of the head are stabilized in relation to the (stationary) sight object. This effect should, first of all, enable airplane pilots to fixate on goals outside rotationally vibrating airplanes. To observe instruments in the cockpit, however, this vestibular controlled eye movement proved to be somewhat disadvantageous.

Resonance of the bulbus, the surrounding tissue, and/or the intraocular structures, including blurring of the picture on the retina during higher vibration frequencies

If the head vibrations reach a certain level, many people experience resonance of the eyeballs in the range 20–100 Hz which, according to Coermann (1938, 1939), at 50 Hz and 2 m/s² acceleration on the seat, for example, should diminish visual acuity to about half. As shown in the section "Vibration Behavior of the Eye" (see p. 44), in which the human and animal experiments of Dupuis and Hartung (1980a) are presented, resonance of the bulbus occurs with a maximum at 20–25 Hz [or at 12.5–25 Hz according to Ohlbaum (1976)]. At the same time, the greatest reductions in visual performance were also found in this frequency range (Figs. 35 and 37). This has also been established in crew members of certain helicopters (Hartung 1983). Oscillation of the image on the retina is also believed to cause a reduction in visual acuity, seen as a "blurring" of vision.

Resonance in the visual cortex

Tachistoscopic testing is not only useful for vision but also for concentration in general. From the experimental results on visual reaction time, it is clear that under vibration stress there is increasing deterioration in performance. The rapid perception of differentiated structures (numbers) especially is considerably reduced. This may be caused by the appearance of resonance in the visual cortex. At any rate, Schoenberger (1974) believed he had been able to show experimentally that during vibration stress only the peripheral processes like information input and output are affected, not the central processing of information.

Furthermore, it must be assumed that vibration-conditioned, psychophysical false sensations and reduction in motivation and concentration can also lead to reduced visual perception.

A description of the cortical excitation condition is possible via flicker-fusion frequency (FFF). Thus, the temporal resolving power of the human eye declines when light flashes quickly follow one another, according to Müller-Limmroth (1961), during physical and mental work. The reduction in FFF is not simply dependent on fatigue, however, as is maintained by a few American psychologists, and does not progress in parallel with the degree of fatigue. Instead, the conditions in the eye are more complicated. It appears to be certain that, above all, central nervous factors have an effect on FFF. On the basis of an extensive literature review, Müller-Limmroth (1959) has pin-pointed many kinds of factors that influence FFF.

Under sinusoidal vibration stress with frequencies between 1 and 10 Hz, Dupuis (1969) established changes in FFF, which correlated rather well the biodynamic vibration behavior of the trunk and subjective sensation. In addition, Dupuis et al. (1974) found a significant decrease in FFF at about 1 Hz during 45-min exposure to random truck vibration with K values $41-82$ ($a_{wz} = 2.05-4.1$ m/s^2). The results of Christ (1973) were similar. He investigated the influence of rest pauses during vibration exposure, using random tractor vibration, with K = 27.5 ($a_{wz} = 1.37$ m/s^2). Under vibration stress, FFF decreased constantly with increasing exposure time but the pattern was interrupted during each interval (Fig. 38).

If one summarizes the current information on the effects of mechanical vibration on sensory functions, there is no clear answer whether impairment in hearing occurs or not. During rotational vibration, the equilibrium organ is clearly involved, though. Kinetosis can be caused by low-frequency vibration, but there are still gaps in our knowledge, especially in connection with the kind

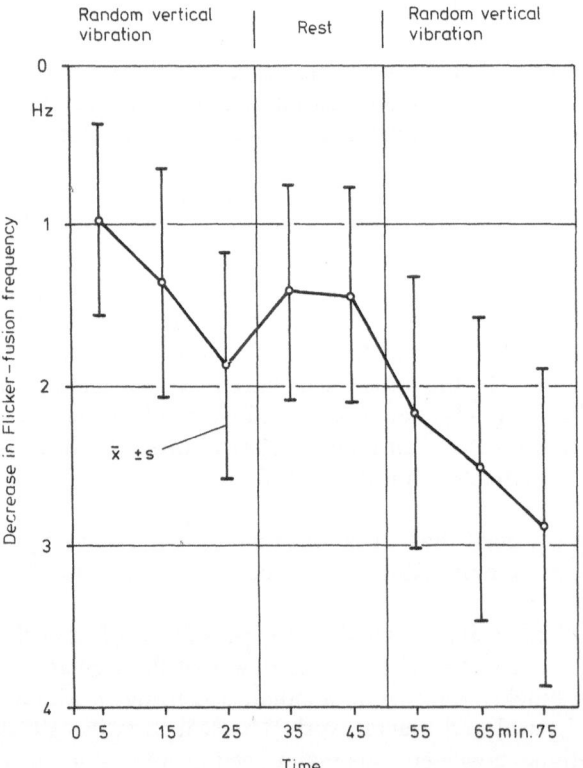

Fig. 38. Reduction in flicker-fusion frequency (FFF) under interrupted random vibration with rms-weighted acceleration $a_{wz} = 1.55$ m/s^2 (Christ 1973)

of vibration stress, the synergistic effect of stress factors, and the adaptation process. The influences on visual perception have been investigated rather thoroughly, but because of the many relevant influencing factors, the genesis of these influences is not yet completely known. With regard to visual perception times, particularly great impairment has been found.

5.4 Sensorimotor Performance

As long as biodynamic and physiological reactions occur, particularly regarding changes in sensory perception, then the sensorimotor performance ability will also be impaired. The extent of such influences will, however, depend on the physical factors of the vibration stress, as well as on the type of sensorimotor task and individual requirements such as motivation. As far as impairment of performance is concerned, a distinction should be made between the direct mechanical effects of vibration on the hand-arm or foot-leg control system and the indirect psychophysiological effects of vibration stress. The former starts immediately with the onset of vibration and stops when vibration stops. In contrast, the latter first develops slowly and sometimes still endures for a period of time after the end of vibration (Coermann 1962). Another time-dependent trend can supervene, depending on the development of practice for the particular task.

Direct mechanical influences can be similar between subjects, whereas the psychophysical influences can evoke strong scatter of performance data. Rühmann (1978) believes that reduction in performance can be ascribed primarily to the direct effects on the processes of visual reception of information and motor reaction to this information. More than 100 experiments have been published on vibration-caused impairment of performance. The most complete review with regard to hand-arm control has been assembled by Lewis and Griffin (1978). However, other important works should also be cited (Dupuis et al. 1955; Schmitz 1959; Schmitz and Simons 1959; Hornick et al. 1961; Fraser et al. 1961; Coermann et al. 1962; Huang and Suggs 1965; Guignard 1965; Harris and Shoenberger 1966; Holland 1967; Sjøflot and Suggs 1973; Pleszczynski et al. 1973; Christ 1973; Dupuis et al. 1974; Gray et al. 1976; Lewis and Griffin 1976; Cohen et al. 1977; Rühmann 1978; Lewis and Griffin 1979b; Stave 1979; Gauthier et al. 1981).

It should be mentioned that almost all experiments have been conducted under laboratory conditions with simulated performance tasks, as field experiments are generally accompanied by additional stress factors and are usually not reproducible. The sensorimotor tasks used were mainly steering or tracking tasks with hand operation − seldom with foot operation, reaction tests, or tests on constancy of foot pressure. Stave (1979) also carried out very realistic flight-simulation experiments with pilots who were exposed to vibration at frequencies of $6-12$ Hz and accelerations of $1.0-2.7$ m/s² (at the vertebral column). It is interesting that under this vibration stress, steering performance (flight deviation) was not reduced but improved. This improvement can be attributed to the high motivation and efficiency of the pilots. In 6% of the tasks, however, the pilots made short-term sudden mistakes, which could often be unclarified and connected with so-called pilot-error accidents.

As a result of the many varieties of vibration stress parameters, and the various performance requirements, it is difficult to derive generally valid data from the literature; however, a few tendencies can be pointed out. Concerning the influence of vibration intensity, Christ used 2-h random vibration stress (simulated tractor vibration) in his experiments, with a frequency-weighted rms value of acceleration 1.55 m/s² (K = 31, according to VDI 2057, 1981). He was not able to establish impairment of performance in a tracking task with a steering wheel. Under 6-h stress from random vibration with rms-acceleration 1.6 m/s², no decrease in performance in the tracking tasks carried out by cadets was observed (Holland 1967).

On the other hand, numerous experiments have shown a drop in sensorimotor performance with increasing vibration intensity in the same frequency spectrum (Schmitz 1959; Fraser et al. 1961; Harris and Shoenberger 1966; Lewis and Griffin 1976). Buckout (1964) found a significant decline in performance during 5-Hz vibration with rms acceleration 1.8 m/s², and Dupuis et al. (1974) during simulated truck vibration with rms acceleration 2.0 m/s². Gray et al. (1976) came to the conclusion in their experiments that short-term stress of 1- to 4-min duration, with an rms acceleration of 2.8 m/s² at 5 Hz, leads to a noticeable reduction in performance. Regarding the influence of vibration frequency, Coermann et al. (1962) showed particularly great impairment of performance between 3 and 12 Hz. The results were more concrete in the experiments by Schmitz (1959), Hornick (1962), Buckout (1964), and Lewis and Griffin (1979), with the greatest effects at around 4−5 Hz. Although these data are valid for the vertical vibration direction, z, the greatest impairment can be expected in the horizontal vibration directions, x and y, at frequencies of 1−3 Hz (Lewis and Griffin 1978). Harris and Shoenberger (1966) have proved that impairment in performance has a tendency to follow the general frequency-dependent curves of physiological tolerance, i.e., above the resonance frequencies, performance will improve with increasing frequency.

Fraser et al. (1961) have examined the effects of the vibration direction, inasmuch as the sensorimotor steering performance is mostly influenced by vibration in the vertical direction z, less by vibration in the horizontal direction y, and not at all by vibration in the horizontal direction x. Huang and Suggs (1965) found equally strong effects, using vibration stress in the x and the y directions. In general, the greatest impairment is found when the vibration direction and control direction are the same.

With respect to the influence of the form of vibration, Bennet et al. (1976) have proved that when the root-mean-square value of the vertical acceleration is the same, 2.1−3.5 m/s², steering performance is reduced more during vibration than during sinusoidal vibration.

The data published on the influence of the duration of exposure vary a great deal, probably because the vibration-stress test conditions used, as well as the experimental subjects and their motivation, have differed so much. Lewis and

Griffin (1978) concluded that there is little to base ISO guideline 2631 time-dependency curves on regarding reduction in performance ability for control tasks. Gray et al. (1976) concur with this conclusion. In addition, minor vibration stresses occurring over several hours seem to have a "waking effect" which counteracts the fatigue effects by means of activation of the reticular formation. Thus, Christ (1973) was not able to prove any significant differences in performance between continual vibration stress of 2-h duration and interrupted vibration stress, also of 2-h duration. If exposure to vibration lasts longer, then monotony could again lead to reduction in performance.

There is no doubt that the layout of controls (displacement, control force, direction of movement) and their position have an influence on impairment of performance by vibration. Thus Dupuis et al. (1955) were able to optimize the accelerator for agricultural tractors according to operating displacement, force, and dimensions, as well as adaptation to the anatomical mobility of the ankle, so that control errors could be kept relatively slight under exposure to vibration. Rühmann (1978) has conducted research on hand-operated controls, with variations in construction, for use when under the influence of random roll vibration. Pleszczyinski et al. (1973) proved that vibration-induced relative vertical movements of 150 and 200 mm between the driver's seat and the steering wheel significantly increase steering errors.

Except for the tracking tasks already mentioned, the influence of vibration on other sensorimotor tasks, such as the mounting tasks and precision engineering activities, has not been investigated. Summarizing the results from the literature, it can be stated that during low-vibration stress, the effects of fatigue can be compensated for by corresponding motivation. With increasing vibration intensity, however, a decrease in performance must be expected. Absolute limit values, under which no impairment is to be expected, cannot be given because of the variety of tasks, the construction of controls, and individual requirements. When under vibration stress resulting in high transmission to the human body, especially in the resonance regions of the body parts, there is particularly great impairment in performance. If possible, the control movement and vibration movement should not follow the same direction. Under practical working conditions, one should always be aware that, in addition to vibration stress, other stress factors can also influence performance.

5.5 Subjective Perception

So far, the biodynamic vibration behavior of the human body, the changes in physiological functions, and the potential influences on sensorimotor performance have been discussed regarding exposure to mechanical vibration. They can be regarded as "objective" effects insofar as they can be objectively measured. However, as regards the extent of human strain, the subjective sensation, i.e., the degree of annoyance of this vibration, is also important. One must use the term "subjective" because the person doing the evaluating applies an individual, subjectively perceived scale that can differ from the scales of other people. The reason for this is that an objective, quantitative scale cannot be established for vibration sensation.

5.5.1 Vibration Perception and Vibration Sensation Memory

Subjective evaluation of vibration has a manifold, practical significance. For example, people who have been exposed to high mechanical vibration can often trace health problems back to the vibration stress. This evaluation can be, but is not necessarily objective. On the other hand, it is also possible that strong perception of vibration with a subjectively high degree of annoyance leads to psychophysical reactions that are then perceived as objective. Finally, one occasionally has to rely on the subjective judgment of test subjects when, for some reason, the objective physiological and psychological criteria cannot be measured in laboratory or field experiments.

One of the most important reasons for taking the subjective degree of the vibration sensation into consideration is, however, to adjust technology to both the objective and subjective requirements of man. This is the goal of every ergonomic design.

In order to make the right choice of experimental method, it is necessary to consider the potential influences and basic limitations of the evidence of subjective measurements (Dupuis 1969):

1. As already presented in the section "Biological Prevention and Control Mechanisms Against Mechanical Vibration" (see p. 12), there are no special and exclusive vibration receptors in the human body. Instead, a variety of perceivable effects can occur in the different body regions, which are evaluated differently by each individual. Thus, in our own series of experiments it was established that there are groups of persons who are especially sensitive in the region of the head and others more sensitive in the region of the internal organs of the abdomen. Although undoubtedly the evaluation given is partly determined by this factor, the experimental subjects should be instructed to render an evaluation of the sensation in all parts of the body that is as comprehensive as possible.

2. Man has only a very short vibration memory, which is even worse for the random type than for the periodic type. This memory can only be trained to a certain degree by very careful observation and memorization on the part of the subject. When two different vibration processes are to be compared, they must be presented immediately after one another. During the single experiments in which a subjective judgment is demanded from the subject, the experiment must, first of all, not be too short in order to leave a sufficient impression of the vibration effect; secondly, the experiment must also not last too long so as not to overtax the vibration memory. This method has proved to be empirically successful when sinusoidal vibration was used if the length of the experiment was chosen to be between 15 and 30 s (a somewhat longer time for random vibration). The pause necessary to switch over should not exceed 10–15 s when two experiments are being conducted for evaluation.

3. The rating of subjective sensation according to an evaluation scale (e.g., evaluation concepts like "strong," "unpleasant," and "scarcely tolerable," or by grade) presents difficulties if fine differentiation of vibration stress is involved. The results obtained are more reliable, according to Dieckmann (1963), if only comparative statements are made (e.g., "equal," "stronger," "weaker") in comparison with vibration stress experienced previously. If one completely dispenses with such graded statements and instead permits the subjects simply to adjust the vibration intensity continously to correspond with the sensation quantity or the degree of annoyance, then in our experience the precision of the results can be improved.
4. Because of the difficulties in verifying subjective judgments of mechanical vibration, as already discussed, other stress factors, such as noise, climatic conditions, or body posture, should be strictly eliminated or controlled.

5.5.2 Vibration Perception in the Sitting and Standing Postures

It is interesting that most experiments on subjective evaluation of vibration deal with the extreme boundaries, that is, with the perception threshold or the short-term tolerance limit. The is probably because the criteria for these boundaries are easy to describe. Although knowledge of such boundaries is also important, curves of equal sensation of vibrations between these boundary values are important for the evaluation of vibration stress encountered in everyday life.

Mallock (1902) was apparently the first to attempt to understand the subjective annoyance of mechanical vibration. He investigated the complaints of inhabitants in the neighborhood of London's Hyde Park, who were disturbed by the shaking caused by traffic. In the Federal Republic of Germany, Reiher and Meister (1931) presented formulas on subjective vibration sensation for the first time. Their formulas were based on numerous experiments using a large number of subjects. There are more than 100 publications about experiments on subjective vibration perception as related to physical parameters. However, only the following need to be mentioned here (Reiher and Meister 1931; Zand 1931; Constant 1932; Janeway 1949; Jacklin 1936; von Bekesy 1939; Helberg and Sperlin 1941; Postlethwaite 1944; Best 1945; Goldmann 1948; Zeller 1949; Mc-Farland 1953; Helling 1964; Miwa 1967, 1968b; Soliman 1968; Dupuis 1969; Lange 1971, 1974a; Dupuis et al. 1972; Splittgerber 1972; Fothergill and Griffin 1977b; Whitman and Griffin 1978; Oborne et al. 1981).

Unfortunately, despite the merits of the results of the authors, the experiments carried out during the 1930s and 1940s are only usable to a limited degree. This is because the measurement technology at that time (e.g., the size and weight of the instruments measuring acceleration) was limited and resulted in considerable measuring errors.

Dieckmann (1957) has dealt with the publications that had appeared up to that time (Reiher and Meister 1931; Zeller 1949; DIN 4150 1939 ≐ Pal scale; Janeway 1949; Jacklin 1936) and their formulas or concepts for the evaluation of vibration perception. Using his own research results, he first defined a

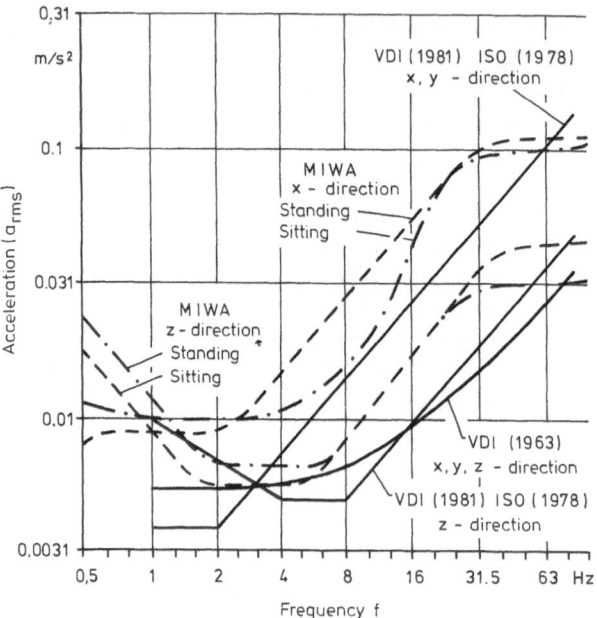

Fig. 39. Thresholds of the frequency-dependent perception of mechanical vibration in the x and z direction in sitting and standing body postures (Miwa 1967; VDI 2057 1963, 1981; ISO 2631 1974)

"K value" that was later included in VDI recommendation 2057 (1963) as the "perception quantity." The K-value scale was chosen so that the threshold of vibration perception would correspond to a K value of about 0.1. This lies at about 0.006 m/s² (rms) between 0.5 and 5.0 Hz, to climb then as the frequency increases further (Fig. 39). Revised frequency-dependent curves of the threshold of vibration perception, differentiated for the x, y, and z directions, were included in ISO 2631 (1974) and in revised VDI guideline 2057 (1981). Experiments to determine the perception threshold data exhibit considerable scatter, and one can only speak of a "perception-threshold range." Splittgerber (1972) has determined the perception threshold in up to 16 subjects in different body postures and in 67 subjects when sitting in an office, the floor of which was stimulated with sinusoidal vibration (13 and 23 Hz). In this way it was possible to obtain an approximate normal distribution of threshold values. No sex-dependent differences were established. The results of the experiments of Miwa and Splittgerber roughly correspond with a threshold curve of K = 0.1, according to VDI 2057 (1963).

Although the threshold values are of interest, especially with regard to the evaluation of vibration stress in apartment buildings, the higher vibration stress in a variety of places of work is significant. In this connection, frequency-dependent curves of equal perception on a higher acceleration level are important.

Using an electrohydraulic simulator, which enabled comparative experiments with a "standard" or "reference" vibration to be carried out, Dupuis

(1969) obtained frequency-dependent curves of equal vibration perception. He used 17 subjects in the sitting position and exposed them to vertical vibration z at three intensity levels. Comparison of the statistical scatter and the mean-value curves showed that the intraindividual variation coefficient (the mean error in repetition for one subject) amounted to 20% and that of the interindividual variation coefficient 42%. Later, Dupuis et al. (1972) further supplemented these curves by means of two more detailed descriptions, "easily perceived" and "30-min tolerance boundary" (Fig. 40). The curves contributed towards frequency evaluation curves according to ISO 2631 (1974) and VDI 2057 (1981). The curve features have distinct troughs around 5 Hz, which come from the main natural frequency of the human body in the z direction and the resulting increased sensitivity at this frequency. Above 5 Hz, the values clearly increase with increasing frequency. Below 5 Hz, the values increase up to 2−3 Hz. However, as the frequency decreases further below 2 Hz, the curves drop again. This decline could indicate the appearance of kinetosis at the upper tolerance boundary which, however, is only of practical significance below 0.5 Hz (see section "Kinetosis," p. 66).

At low-vibration stress levels, Dempsey et al. (1979) have found somewhat flatter curves of equal vibration perception compared with those at higher vibration stress (Fig. 40).

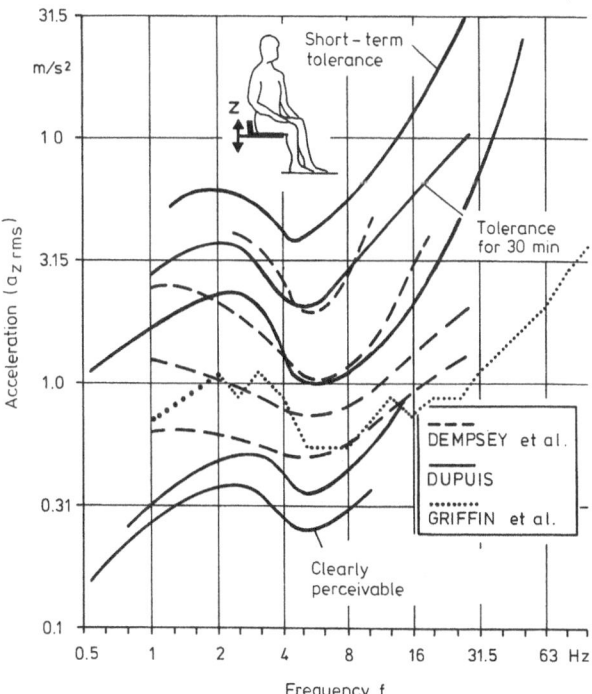

Fig. 40. Frequency-dependent curves of equally strong perception of mechanical vibration in the vertical direction z in the seated position (Dupuis 1969; Dupuis et al. 1972; Dempsey et al. 1979)

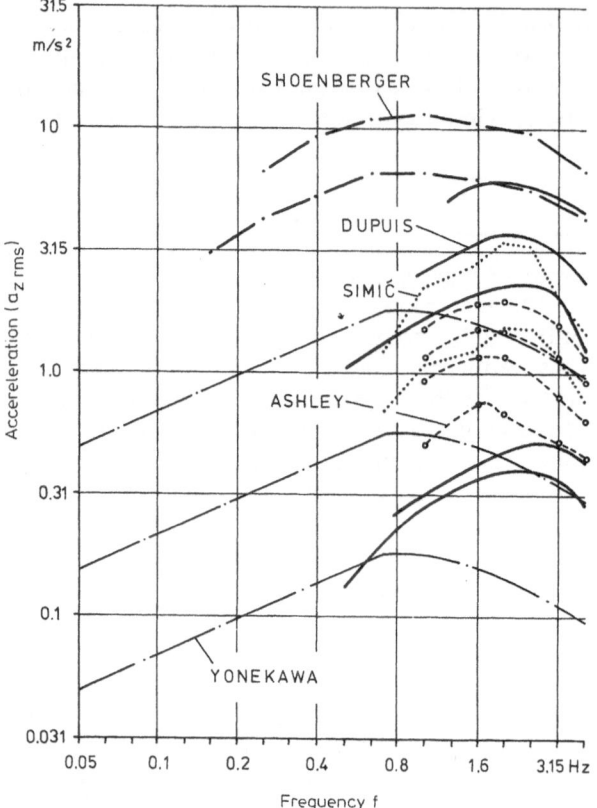

Fig. 41. Curves of equally strong perception of mechanical vibration at a frequency range of under 4 Hz in the vertical direction z while sitting (Dupuis 1969; Ashley 1970; Simic 1970; Dupuis et al. 1972; Yonekawa and Miwa 1972; Shoenberger 1975)

The tendency for the decline below 2 Hz, which is also visible at lower acceleration intensities and thus indicates a second region of special sensitivity in man, is already evident in the results of Parks (1962), Ashley (1970), Simic (1970), Yonekawa and Miwa (1972), and Schoenberger (1975) (see Fig. 41). Yonekawa and Miwa (1972) have shown that this tendency is also valid for vibration stimuli in the vertical and horizontal directions and for the three body postures, standing, sitting, and reclining. Dupuis et al. (1972) have also verified these findings for the sitting position, using combinations of sinusoidal vibration.

It is difficult to obtain consistent data for the upper tolerance limit for mechanical vibration, because the physical constitution of the subject plays a role, as does the "tolerance" criterion. In this connection, the duration of vibration stress must, above all, be taken into consideration. Fothergill and Griffin (1977b) presented root-mean-square values of vertical acceleration described as "very uncomfortable" in 10-Hz vibration experiments, each of 5 s duration, using a scaled evaluation of subjective perception. Their paper also gives the "very uncomfortable" level of other authors:

Fothergill (1972)	2.5 m/s²
Jones and Saunders (1974)	3.7 m/s²
Oborne and Clarke (1974)	2.3 m/s²
Fothergill and Griffin (1977 b)	2.7 m/s²

The corresponding value for 3-min tolerance given by Miwa (1968 b) was 3.2 m/s². If one compares these values for 10 Hz vibration with the curves of Dupuis et al. (1972) in Fig. 40, it is apparent that they are similar to those for 30-min tolerance. There is apparently no clear demarcation between the criteria "very uncomfortable" and "tolerance for 30 min." Magid et al. (1960) had already determined these frequency- and exposure-dependent tolerance curves in earlier studies. The acceleration level for the tolerance curves in the frequency range of 5−7 Hz corresponded to 3-min duration at about 5 m/s², and to 1-min duration at about 10 m/s². The subjects − probably soldiers − must have had to show considerable heroism, however, because they would have bounced on their seats at 9.8 m/s² acceleration. The tolerance curves given by Miwa (1968 b) correspond very well with the curves of ISO 2631 (1974) guideline as regards frequency dependency.

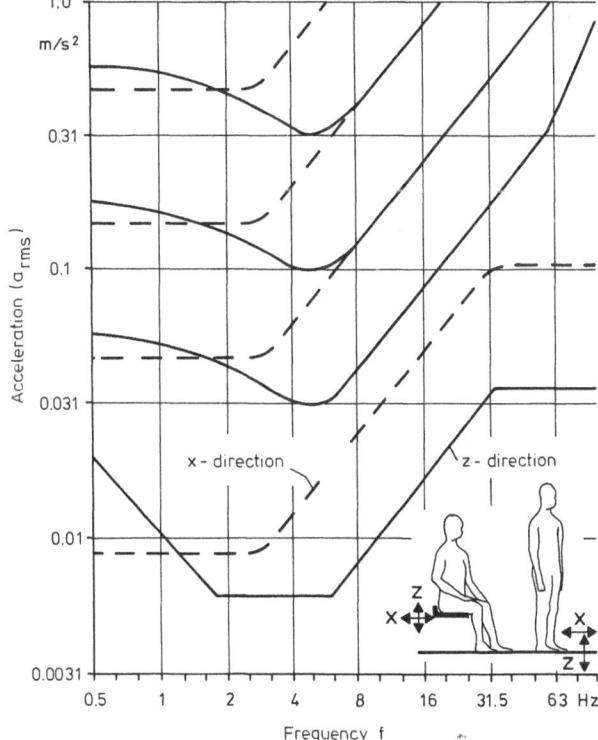

Fig. 42. Frequency-dependent curves of equally strong perception of mechanical vibration in vibration directions x (horizontal) and z (vertical) while sitting (Miwa 1967)

The experiments by Miwa (1967), discussed above, which were carried out under the influence of vibration in different directions, defined the perception threshold. For the seated person there are also corresponding frequency-dependent curves of equal vibration perception in the x and z directions at different acceleration levels (Fig. 42). It is apparent that the data for the two vibration directions differ, above all, with regard to the frequencies where the curve is at a constant acceleration level. This arises from the biodynamic vibration behavior of the human body, which is discussed in the section "Vibration Behavior in Sitting Posture" (see p. 26). From this it is apparent that tolerance to horizontal vibration is higher in comparison to vertical vibration above about 3 Hz, as is shown in Fig. 42. These results are also valid for the y direction because, according to Miwa, there is no difference in perception between the x and y directions.

5.5.3 Vibration Perception in Reclining Posture

The frequency-dependent curves of equal vibration perception differ only slightly for the sitting and standing position (as shown by Miwa's data in Fig. 39). However, for the reclining posture one can expect another form of frequency-dependent perception because of the great differences in biodynamic vibration. There are a number of experimental results pertinent to this question (Temple et al. 1964; Seris 1969; Miwa and Yonekawa 1969; Helling (−); Szameitat 1976; Szameitat and Dupuis 1976; Ashley 1978; Dupuis and Hartung 1981).

To obtain basic, comparable results on vibration perception for such experiments requires the vibration platform to be a rigid support base for the body, in the same way as in experiments in the seated and standing position. The results of experiments that have been carried out under such constraints are presented in Fig. 43 for the vertical vibration direction, x, and in Fig. 44 for the horizontal direction, z. The results for the horizontal direction, y, can be disregarded, as these correlate with those for the z direction.

For vertical vibration, the experiments carried out at various acceleration levels show quite good correlation between authors (Fig. 43). Up to about 3 Hz, the perception curves progress at approximately constant acceleration. Between 3 and 7 Hz the curves fall, and between 7 and 63 Hz they are again at a constant (although reduced) acceleration level. The resonance behavior of various parts of the body also follows this tendency (see section "Vibration Behavior in Reclining Posture," p. 23). In the curve region around 63 Hz, resonance occurs, above all, in the skull.

Under horizontal vibration stress, curves of equal perception show the relatively highest sensitivity to be about 2−4 Hz, based on the appearance of resonance, and the curves climb with increasing frequency (Szameitat 1976). This also holds true for the threshold curves given by Miwa and Yonekawa (1969). The curves given for the reclining position do not appreciably differ from those for the sitting and standing position.

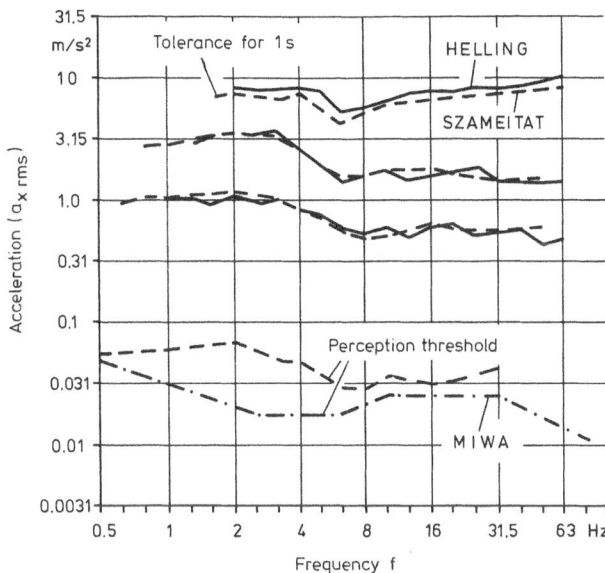

Fig. 43. Frequency-dependent curves of equally strong perception of mechanical vibration in the vertical direction x while reclining (Miwa and Yonekawa 1969; Szameitat 1976; Helling 1978)

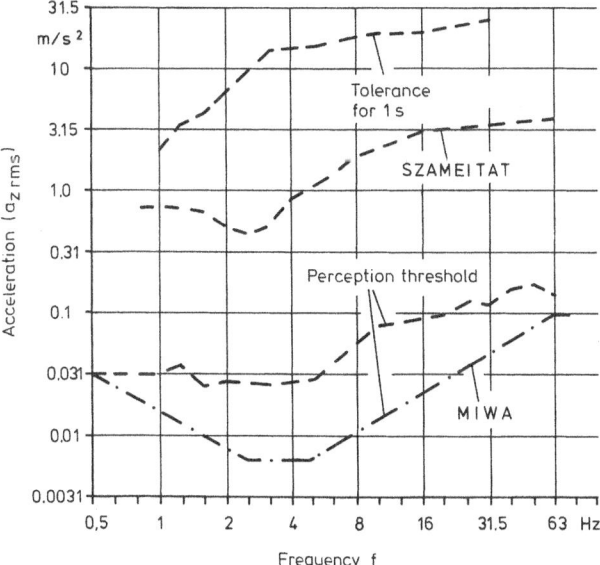

Fig. 44. Frequency-dependent curves of equally strong perception of mechanical vibration in the horizontal direction z while reclining (Miwa and Yonekawa 1969; Szameitat 1976)

Finally, it should be mentioned that Helling (−) has established frequency-dependent perception curves for rotational vibration, in which the horizontal rotation axis in the y direction lies under the center of gravity of the reclining person. Under these conditions, he found the greatest vibration sensitivity at about 30 Hz.

The experimental results of Szameitat (1976), which correspond closely to those of the other experiments described, led to a draft standard which is included in VDI recommendation 2057, Part 2 (1981).

Helling (−) and Szameitat (1976) investigated the use of soft foam-rubber pillows. Ashley (1978) used NATO stretchers and Dupuis and Hartung (1981) tested vacuum mattresses. The use of such experimental variations should permit certain questions to be answered.

In this way it was shown that suspended stretcher coverings and soft pillows for the head can reduce sensitivity by decreasing the transmission of high-frequency vibration to the head. On the other hand, vacuum mattresses, which become very hard because of the air evacuation, do not reduce vibration sensitivity at higher frequencies. At frequencies below 3 Hz, however, these vacuum mattresses effect relative body stabilization (immobilization) in the horizontal vibration directions and avoid strong resonance behavior in this way. Therefore, the perception curves climb. As a result of the immobilization effect, vacuum mattresses are especially suitable for transport of the severely injured.

In summary, it is apparent that there is enough current knowledge on subjective vibration perception to provide us with sufficient information on the influence of body posture, vibration frequency, and acceleration intensity. These results, presented in connection with the current knowledge of biodynamic vibration behavior, changes in physiological reaction, and chronic effects, are the essential bases for the evaluation of vibration (see section "Laws, Regulations, Standards, and Guidelines for the Protection of Man Against Mechanical Vibration," p. 133).

6. Chronic Effects of Whole-Body Vibration

The reactions of the body to stress from mechanical vibration, presented in the section "Acute Effects of Mechanical Vibration" (see p. 12), deal with individual organs and functions. It is now time for the following question to be asked: Are these physiological or pathophysiological changes in reality the external symptoms of a vibration-induced illness of the whole organism? Above all, some Soviet scientists are of this opinion. Andreeva-Galanina (1967) has especially called this concept "vibration illness" and has used it in the sense that the totality of vibration-induced changes in the human organism can be characterized (Rublack 1978). In this systemic "vibration illness," Andreeva-Galanina differentiated between two different forms, which were to be characterized by the following symptoms:

1. Vessel spasms with sensitivity disturbances, particularly the sensation of vibration, trophic as well as central nervous disturbances (evoked by vibration > 35 Hz).
2. Vessel atonia, hemostasis, bone and joint changes (including the vertebral column), minor perception disorders (evoked by vibration < 35 Hz).

The term "vibration illness" is used by Andreeva-Galanina to mean the consequences of all kinds of vibration effects, although she uses it mostly in illnesses resulting from hand-arm vibration. A few Japanese scientists also follow this school of thought, although other scientists do not agree. With regard to this hypothesis, in the spring of 1983 a committee of specialists met in London, by invitation of the Permanent Commission and International Association on Occupational Health. The overall result was that chronic injuries to the central nervous system due to long-term occupational stress from hand-arm vibration had so far not been proven to be generally valid (Gemne and Taylor 1984).

According to current knowledge in occupational medicine regarding the effects of whole-body vibration discussed in this book, a "vibration illness" in the sense of a systemic illness of the total organism has even less validity. However, the chronic effects of whole-body vibration on various organs established so far will be discussed in the following pages.

6.1 Methodological Problems

As required in § 3, section 1, No. 2, of the German law concerning company physicians, safety engineers, and other specialists for occupational safety – the

Occupational Safety Law (ASiG) – effective 12 December 1973 (BGBl. 1 p. 1885), the evaluation of work-related illnesses runs into difficulties when the effect of whole-body vibration at work is considered. Above all, evaluation of such "vibration-induced" damages to health is difficult when the question of the causal relationship between vibration stress and existing damage to health is discussed. Thus, in Germany work-related illnesses resulting from whole-body vibration can only find recognition as occupational illnesses if they fulfill the requirements of § 551, section 2, of the safety regulations of the Federal Republic of Germany FGR National Insurance Regulations (Reichsversiche-rungsordnung, RVO). This means that, according to the current knowledge of medical science, the illness is caused by special effects, to which certain groups of persons are considerably more exposed because of their work than the remaining population.

This evidence can only be collected by means of extensive epidemiological studies (retrospective studies, transversal and longitudinal studies, general morbidity studies) on occupation groups exposed to high stress from whole-body vibration. For this reason, comparative evaluations have to be based on clinical findings and radiological examinations, as well as precise occupational case histories. The people who are initially examined should be reexamined, using the same methods, after several years of professional exposure, in order to discover possible changes in their state of health.

Furthermore, a comparable group of subjects who have not been exposed (control group) should be compared to a group of vibration-exposed workers, using the same methods of examination. Lange (1981) points out the problems arising from the lack of suitable control groups for epidemiological studies in occupational medicine.

Another problem is making a distinction between the people exposed to a low level and those to a high level of vibration stress in such comparative studies. Lange advises that extreme groups with very low and very high stress be compared with one another and that the intermediate groups be omitted.

It should be noted here that most of the epidemiological research carried out on this problem does not, on the whole, fulfill the requirements expressed here. For this reason, the corresponding hypotheses often cannot be answered from the strict scientific point of view and tendencies can only be deducted.

6.2 Work-Related Diseases in National and International Regulations

As a result of the statements in the above section, with regard to the causal relationships, injuries to health that could be related to whole-body vibration cannot be proven in a simple way. This is particularly true for the region of the spinal column, since a high percentage of the total population suffers from defects in the vertebral column anyway.

In contrast to this, the physical effects seen in symptomatology, clinical medicine, pathology, and physiopathology, which are encountered in the hand-arm system as a result of exposure to mechanical vibration, are well known (Dupuis 1982; Laarmann 1977; Klosterkötter 1974). The resulting two occupational illnesses for which compensation is possible are included in the German list of occupational illnesses (Berufskrankheiten-Verordnung, BeKV) of 8 December 1976 (BGBl. I p. 3329; Wagner and Zerlett 1982).

One disease listed deals with "illnesses from vibration in work with compressed air tools, as well as tools or machinery with similar effects" (occupational disease no. 2103). This illness manifests itself as a degenerative injury or disturbance of nutrition resulting from mechanically clogged vessels in the hand, elbow, and shoulder joints (Laarmann 1977; Steinhäuser and Bolt 1979). Examples are arthrosis and aseptic necrosis of the os lunatum, as well as fatigue injuries and fatigue fractures of the os scaphoideum, possibly followed by pseudoformation of arthrosis of the skeletal parts involved. They are caused by work tools and machines with recoil vibration, above all of the low-frequency type (compressed air hammers, electric hammer drills, etc.).

The other occupational disease requiring compensation is "vibration-induced disturbances in blood flow in the hands," (occupational disease no. 2104). In contrast to the first occupational disease, this one is caused by the effect of predominantly high vibration frequencies; it is also called "vibration-induced white-fingers VWF" or "vibration-induced vasospastic syndrome" (VVS). Among other factors, it may be caused by the use of chain saws, electric grinders, impact wrenches, and compressed air hammers (Klosterkötter 1974; Dupuis 1982).

In contrast to the illnesses affecting the hand-arm system, occupational diseases occurring after long-term exposure to whole-body vibration have not been included in the list of occupational diseases of the Federal Republic of Germany. The only published case history relating to a work-caused fracture of the spinal process of the seventh cervical vertebra (so-called clay-shoveller's fracture, occupational illness no. 2107) is that of Zerlett (1963). This case concerned a driver of an earth-moving machine in a brown-coal open-cast mine whose injuries occurred as the result of many years of stress from whole-body vibration (see Fig. 49).

However, the so-called general system of occupational diseases of the Federal Republic of Germany (an enumeration of occupational illnesses in a list of occupational diseases and the so-called general clause, according to § 551, section 2 of the National Insurance regulation, RVO) leaves the possibility open for also recognizing an injury to health as an occupational disease. This may be the case if this illness is not included in the list of occupational diseases, insofar as the remaining requirements are fulfilled according to new findings. According to the verdicts of the Federal Social Court, "new findings" are those which had not been known in decrees concerning the occupational diseases' regulations valid at that time and, therefore, could be considered neither positively nor negatively (Wagner and Zerlett 1982). (See also section "Occupational Diseases, p. 133).

According to the regulations on the prevention, reporting, and assessment of occupational illnesses of 26 February 1981 (BGI, No. 12, p. 137) in the German Democratic Republic (GDR), the following illnesses are recognized: "degenerative diseases of the vertebral column (vertebrae, vertebral disk end-plates, spinal process, ligaments, small vertebral joints) from many years of mechanical overstress." This is no. 70 in the list of occupational diseases in the sense of an occupational disease, following whole-body vibration stress, that is entitled to compensation if it results in considerable limitation of function of the locomotive system and cessation of the occupation connected with vibration stress.

The international list of occupational diseases, ILO agreement no. 121 (1980) contains, in the latest revised version, occupational illness no. 23, the following definition: "Illnesses caused by vibration (illnesses of the muscles, the tendons, bones, joints, peripheral vessels or nerves)." In the report of a conference of experts (Geneva, 14–22 January 1980), it is further pointed out that the "experts recognize the significance of illnesses resulting from low-frequency vibration in the drivers of motor-driven machines, tractors, etc. (ailments of the lumbar vertebral column, sciatica, etc.)"; however, they emphasized that the ailments are not typical (specific) and the occupation-related origin is therefore "often difficult to prove." Furthermore, it says that they (the experts) recognize "the occupational origin of diseases of the vertebral column resulting from vibration of the whole body and from shaking, especially in drivers of motor-driven machinery and tractors."

6.3 Vibration Effects on the Skeletal System in Animal Experiments

As the question of potential injury to the spinal column from whole-body vibration cannot be answered experimentally on man, sporadic animal experiments have been carried out in this area. Witt and Fischer (1980) exposed 24 guinea pigs daily for about 2.9 h to vibration of 6 Hz and 1.4 m/s² rms acceleration in the direction of the vertebral column. After exposure for up to 203 h over a period of 75 days, histological examination revealed vibration-induced lesions in the vertebral joints, the paravertebral musculature, and in the blood-filled spaces of the spongiosa. Changes in the vertebral joints are, above all, reported to result in great pain. Homogenized muscle cells with loss of transverse striation were found in the muscle specimens. This phenomenon was ascribed to muscle hypertonia caused by reflexes from the vibration exposure. Greatly extended blood-filled spaces in the spongiosa, with regular pools of blood, were observed and are considered to be scatter-conditioned disturbances in outflow, which are supposed to lead to rarefaction of the spongiosa trabeculae and could also have an arthrosis-triggering effect. Bone exposed to whole-body vibration in animal experiments becomes fragile and brittle as a result of the increase of inorganic components in the constant calcium content. These findings, corresponding to the usual aging process of bones, must thus be regarded as premature structural weakness caused by vibration (Jankovich 1971).

Fassbender (1979) conducted experiments using 32 Wistar rats and a control group to clarify whether changes occur in the skeletal musculature and the joint cartilage as a result of various levels of vibration stress in the low-frequency range (3.5 – 10 Hz). By means of electron-optical and light-optical methods of investigation, he observed that procreation of the mitochondria and growth of giant cells took place in the musculature after exposure to vibration stress for 4 weeks. After 8-week stress, the mitochondria show discrete degenerative stages. However, above all it was shown that long-term vibration can cause cartilage-cell degeneration and death after a transitory phase of disturbance of secretory cell activity. This results in the development of osteoarthrosis.

6.4 Epidemiological Research on Professional Groups Exposed to Vibration

In the last three decades, the following professional groups exposed to stress from whole-body vibration have been studied; some of the research has included extensive epidemiological investigations:

Tractor drivers (agriculture and forestry)
Drivers of earth-moving equipment (excavator, leveler, grader, loader, dozer, etc.)
Drivers of other construction equipment
Loader drivers in underground mines
Drivers of so-called auxiliary equipment in surface mines
Drivers of heavy-duty trucks
Truck drivers
Drivers of pickup trucks
Bus drivers
Streetcar drivers
Railroad personnel
Airplane crews (helicopter and jet pilots)
Ship crews
Cement factory workers on vibrating platforms

No epidemiological studies have been conducted on workers exposed to whole-body vibration in buildings. From the case reports available, analysis of the literature should furnish data on epidemiology and morbidity that would help to clear up the possible connection between whole-body vibration stress and damage to health.

With regard to this question, the international literature has been summarized by Heide (1978); Heide and Seidel (1978); Rublack (1978); and Sandover (1981) and is considered later.

6.4.1 Diseases of the Spinal Column

Of all the disturbances and injuries to health arising from long-term exposure to whole-body vibration that have been observed and described, complaints and diseases of the spinal column take first place. This is understandable if one considers that vertical mechanical vibration at the seat of a sitting person proceeds directly to the vertebral column. The vertebral column, as a sensitive structure, has a particularly strong reaction to such stress (see section "Vibration Behavior of the Spinal Column," p. 31). The results from the available literature on spinal complaints are presented in the following pages, subdivided by professional groups.

6.4.1.1 Tractor Drivers

Fishbein and Salter (1950) have probably conducted the earliest study, in which 378 American orthopedists were questioned in an attempt to shed more light on the connection between stress from driving trucks and tractors and injuries to the vertebral column or to the support apparatus. At that time (1950), either temporarily or daily, more than 10 million Americans professionally drove trucks, agricultural tractors, buses, taxis, locomotives, and construction equipment, not to mention the extremely large number of military vehicle drivers in the armed forces.

One of the questions asked was: "Have you seen patients in whom disorders of the spine and supporting structures might be ascribed to driving in trucks or tractors?" "None" was the answer of 109 orthopedists; 179 answered with "a few," and 54 with "many." The remaining 36 assumed that driving a truck or trailer is an aggravating factor but not, however, a causal factor. The total number of cases given was 7,851. Table 8 presents the responses to the following question: "Which of the following specific disorders do you believe may have been caused in whole or in part by driving trucks or tractors?"

Furthermore, 37 other health problems were listed by the orthopedists queried: e.g., lumbosacral complaints, aggravation of an existing spondylolisthesis condition, muscular tension.

In a larger series of epidemiological studies, Rosegger and Rosegger (1960) and Rosegger (1966, 1970) commented on the morbidity in tractor drivers. The data obtained from work case histories and radiological examinations of the bones and joints of a total of 371 tractor drivers indicated that there were a great number of premature degenerative changes in the vertebral column in this group.

Table 8. Number of positive replies by physicians in connection with truck or tractor vibration and impairment to health (Fishbein and Salter 1958)

Spinal column injuries	25
Disk hernias	95
Osteoarthrosis	103
Spondylosis	41
Subluxation	20
Sacroiliac complaints	72
Traumatic fibrositis	109
Coccygeal pain	82

Pathological changes in the cervical vertebral column were observed in 17.9% of the tractor drivers; 9 of the 57 drivers with injuries to the cervical vertebral column suffered from so-called clay-shovelers' fracture, a fracture of the spinal process in the region of the cervical vertebrae.

In the thoracic and lumbar areas of the spine, they found radiologically confirmable changes in 71.3% of the tractor drivers. The pathological changes established occurred in the thoracic area of the spine alone in 51.6%, the lumbar area alone in 8.1%, and both the thoracic and lumbar regions of the spine were involved in 40.3% of the cases examined.

The changes in the thoracic part of the spine consisted of juvenile kyphosis, scoliosis and kyphoscoliosis, osteochondrosis, and spondylosis deformans, as well as Schmorl's nodes. Changes in the region of the lumbar vertebral column resulted in kyphosis, scoliosis, kyphoscoliosis, osteochondrosis, spondylosis deformans, Schmorl's nodes, vertebral body shifts, arthrosis of the small joints, Baastrup phenomenon, and lumbosacralization.

In the opinion of the authors, the work stress that occurs before the tractor-driving activity, especially during the developmental years (14 – 20 years of age), had apparently no influence on the radiologically confirmable changes in the vertebral column. Nevertheless, it is clear that the pathological findings depend on age and length of exposure (Tables 9 and 10), although they cannot be distinguished statistically.

Table 9. Pathological findings in the spinal column related to age of the tractor driver (Rosegger and Rosegger 1960)

Age group (years)	Incidence of pathological radiological findings (%)
14 17	39.1
18–20	64.7
21–25	70.3
26–30	80.0
31–35	87.5
36–40	80.0
41–50	81.8
Older than 50	87.5

Table 10. Pathological findings in the spinal column related to professional years as tractor driver (Rosegger and Rosegger 1960)

Professional years	Incidence of pathological radiological findings (%)
1	60.4
2	66.7
3	75.5
4	73.6
5	77.8
6–10	77.8
More than 10	80.0

According to this study, the percentage of pathological findings in the vertebral column appears to be lower in tractor drivers who participate in active
sports, in contrast to those who do not (68.4% vs 75.8%). On the other hand, it
was not possible to show that type of constitution has an influence.

The authors concluded that in tractor drivers, the pathological changes established for the thoracic and lumbar parts of the spine outnumber by far the
usual amount of degenerative symptoms. The pathological changes established
in the cervical vertebral column were found to be far higher in younger age
groups than the average for the normal population, as expected from the literature. However, deviations from the normal seem to be less conspicuous.

Lavault (1962) examined 184 sick (unfit for work) agricultural tractor
drivers and determined that in 20% of the workers the complaints were in the
region of the cervical and upper-thoracic parts of the spine, 65.2% in the region
of the thoracic part, and 10.3% in the lumbar and sacral region. In 107 drivers,
of whom 77% were between 17 and 45 years of age, X-ray films were taken of
the spinal column with special attention to the lumbosacral region. An especially high incidence of disk thinning was found in this region (Table 11).

Zimmermann (1964) carried out a large-scale study of 12,874 tractor drivers
in Austria. Using a questionnaire, it was proven that 65% of the drivers queried
complained of problems resulting from whole-body vibration. In 55% of the
drivers questioned, according to the information given by the attending physicians, the disorders were thought to be connected with tractor driving.

From the total subject group, 137 tractor drivers with at least 5 years of professional activity, and who were not older than 40 years were subjected to a
radiological examination of the vertebral column. Of these drivers, 90% had
complaints, which in up to 88% were localized in the back or sacrum. Of those
examined, 80% showed degenerative changes in the spinal column. According
to Junghanns, this percentage is, in the general population, first seen only at the
age of 49. In those under 30 years of age, up to 74% showed these degenerative
changes in the spinal column. The radiological examinations also detected
kyphosis in 46% and scoliosis in 68%.

The distribution of morphological changes was as follows: up to 71% in the
thoracic part of the spine, 4% in the lumbar part, and as high as 25% in both the
thoracic and lumbar regions. The author concluded that these experimental results validly confirm the damaging influence of long-term and recurring vibra-

Table 11. Type and incidence of pathological changes in the spinal column of tractor drivers
(Lavault 1962)

Spinal area	Incidence (%)	Type of pathological change
Cervical spine	2.8	Arthrosis
Cervical spine	4.8	Disk thinning
Lumbosacral area	5.6	Arthrosis
Lumbosacral area	19.8	Disk thinning (of which 15% were at L5/S1 and 6.5% intervertebral disc prolapse)

tion. A specific form of damage to the vertebral column, however, is not known to exist.

In addition, Kubik (1966) believed that the results of his experimental series on 400 tractor drivers indicate that constant vibration has an injurious effect on all structures of the vertebral column.

In a long-term study, Christ (1963) first examined 211 young tractor drivers (average age 17.4 years). After 5 years, 137 of the same drivers were reexamined (Dupuis and Christ 1966b; Christ and Dupuis 1968); after a further 6 years, 106 of the same drivers were examined again (Dupuis and Christ 1972). In the initial examination, postural anomalies were found in about one-third, and in 50% structural disturbances of the spinal column were established radiologically. In the first repeat examination in the series, intensified kyphosis had increased from 34% to 52.6%. In approximately one-third, there was a deterioration in the clinical findings.

Radiologically, apparent deteriorations in the vertebral column were seen in a portion of those examined. For the serious defects, with a definite unfavorable influence (juvenile kyhposis, scoliosis, spondylolysis, spondylolisthesis, pseudospondylolisthesis, retrolisthesis, defects in the vertebral arch and vertebral body, as well as in the articular processes and lumbosacral transitional vertebrae), the condition had deteriorated in comparison with the findings of the initial examination 5 years previously. In 68.7% of the 137 tractor drivers examined, the radiological follow-up revealed serious defects and, in 10.2%, defects with a slightly unfavorable influence were established. Only 21.1% of the tractor drivers reexamined showed no defects in the spinal column.

Of the drivers examined with definitely disadvantageous or slightly unfavorable findings in the spinal column, as demonstrated by radiology, 84.4% complained of problems in the spinal column; however, 74% of those without complaints were shown by radiology to suffer from unfavorable or slightly unfavorable defects in the spinal column. Therefore, no correlation could be established between the radiological findings in the spinal column and the subjective complaints.

Furthermore, the authors established that both the incidence of complaints concerning the spinal column and the pathological findings determined by radiology depend on the length of exposure (tractor-driving hours per year; Fig. 45).

The authors did not risk, however, attributing the deterioration found (i.e., regarding the endogeneous-conditioned changes in the vertebral column) to exterior influences (i.e., whole-body vibration stress) in the tractor drivers after an observation period of 5 years. The number of cases was insufficient and the examination time period too short. There is only the well-founded suspicion that whole-body vibration stress may cause injury to the spinal column.

In the second repeat examination of 106 young tractor drivers of the same group, a further deterioration was determined, in terms of increase in the incidence of significant findings. The results of this long-term experiment are summarized in Table 12.

Nevertheless, the authors do not believe it appropriate to make clear-cut, definitive statements on the causes of damage to the spinal column, especially

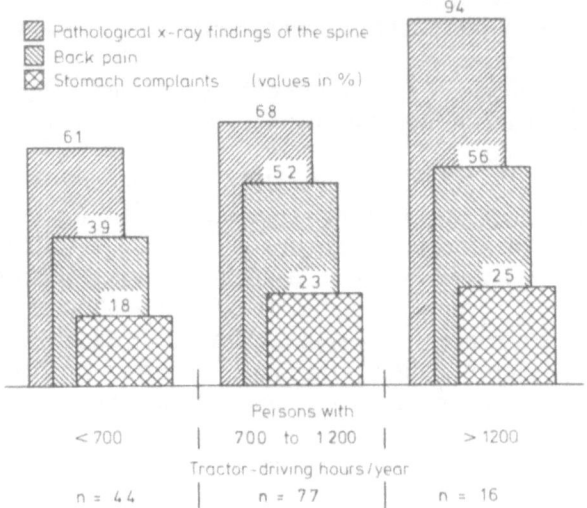

Fig. 45. Incidence of complaints and pathological findings in the spinal column of tractor drivers related to annual length of exposure (Christ and Dupuis 1966 b)

Table 12. Comparison of tractor drivers in a longitudinal study between 1960 and 1977 (various investigations by Dupuis and Christ)

Year of examination	1960–1961	1965–1966	1970–1971
No. of patients examined	211	137	106
Average age (in years)	17.4	23.0	29.3
Radiological findings (in %)			
A. Serious	50.2	68.7	80.1
B. Somewhat unfavorable	22.3	10.2	8.5
C. Probably no influence	14.7	13.1	5.7
D. No pathological findings	12.8	8.0	5.7
Total	100.0	100.0	100.0

with regard to the progressively decreasing number of probands available, after an observation period of 11 years. They concur, however, that the degeneration established in the spinal colum (osteochondrosis, spondylosis, and arthrosis of the vertebral arch joints), which had occurred within the last 5 years in one-fourth of the drivers examined, could possibly be regarded as partly "premature" and "exogenous."

   In a specific experimental series on 60 tractor drivers regarding injuries to health from noise and vibration, as determined by work-case-history questionnaires and clinical symptoms, Seidel and Tröster (1970) showed that of all expressed health disturbances, subjective symptoms in the region of the spinal column took first place. Of the 60 tractor drivers examined, 24 were without complaints, and 25 complained of pain in the area of the vertebral column. The drivers suspected that the vibration stress on the tractor was the decisive factor.

In an epidemiological study, Köhl (1975) surveyed 582 agricultural tractor drivers by means of a questionnaire and found that 61% complained of back pain. The author took X-ray films of the cervical and lumbar spinal regions in 49 tractor drivers of 20−29 years of age. It was established that agricultural drivers exposed to vibration stress had a higher rate of epiphysitis, as well as flattening of the intervertebral spaces, in comparison with a control group of young male nurses of the same age.

Schulze and Polster (1979) discovered occupationally induced injuries to the spinal column in tractor drivers and farmers. In an epidemiological study, 103 tractor drivers who had worked on average for 12 years were compared with a control group of 52 farmers who had worked an average of 19 years. Clinical and radiological examinations established that degenerative diseases of the spinal column had been encountered in the tractor drivers 10−12 years earlier and with a higher incidence than in the average population. In this group the thoracic part of the spine was affected far more than the other spinal parts, with increased exposure time to whole-body vibration stress. In farmers, these degenerative changes in the lumbar area appeared approximately 7 years earlier and in a more pronounced manner, than in the average population. Thus, in this case, the lumbar vertebral column was more strongly affected.

6.4.1.2 Truck and Bus Drivers

Schmidt (1969) conducted a comparative study on 117 healthy, fit-for-work professional drivers of heavy-duty trucks and 95 bank employees, in order to pursue the question of job-induced degeneration of the spinal column. The drivers of heavy-duty trucks, who were exposed to the effects of mechanical vibration and shock, in addition to their job-related hazards, worked sometimes more than 10 h/day as long-distance drivers. They did not load or unload their vehicles. In addition to carrying out extensive, clinical examinations, the author also took X-ray films of the spinal column, particularly lateral views of the cervical spine, and lateral and anterior-posterior views of the thoracic and lumbar spine. The exposed group consisted of people 20−65 years of age whose exposure to vibration stress ranged from 3 to 41 years.

Of the professional drivers, 36 complained of problems in the region of the spine (lumbago, sciatica, lumbar pain, and low back pain).

The following classification was used in the evaluation of the X-ray films:
1. No radiologically detectable morphological change.
2. Early and low-grade osteochondrosis and spondylosis of the spinal column (early, sporadically encountered central border appositions; normal width or not appreciably reduced intervertebral spaces; sporadic surface fractures in the end plates).
3. Medium-grade and distinguishable osteochondrosis and spondylosis of the vertebral column (clear-to-gross border appositions of the vertebral column, above all ventrally and laterally; intervertebral spaces distinctly flattened; marked sclerosis of the end plates; narrowing of the intervertebral foramina).

Table 13. Radiological findings in the spinal column in truck drivers and bank employees (Schmidt 1969)

Findings (in %)	Truck drivers	Bank employees
None	20.5	39.0
Pathological findings in the spinal column	79.5	61.1
Of these:		
Slight changes	36.8	42.2
Moderate-to-pronounced changes	42.7	19.9
Section of spine affected in moderate-to-pronounced changes		
Cervical spine	8.8	8.1
Thoracic spine	26.7	9.1
Lumbar spine	16.2	8.1

The following postural anomalies were found in the 93 drivers with radiological findings: 25 cases of scoliosis, 22 cases of kyphosis, 10 flat backs, 2 curved backs.

In Table 13, radiological findings in the truck drivers are compared with those of bank employees. The vertebral sections involved are also shown.

Distinct changes in the vertebral column have a significantly higher incidence in the driver group of heavy-duty trucks than in the control group of bank employees. The author came to the conclusion that the drivers are prone to premature and more pronounced deterioration of the vertebral column, especially because of the osteochondrosis and spondylosis in the group 31−51 years of age exposed to vibration, as opposed to the control group.

In about 72% of 94 drivers of small delivery trucks who were examined, Kristen et al. (1981) observed pain in the small of the back. In 83%, pathological processes were found in the spinal column (spondylosis, spondylarthrosis, osteochondrosis). The authors regard the degenerative changes in the vertebral column as an adaptation mechanism to continuing high stress.

In a radiological-epidemiological experiment with 213 drivers of a bus company, Barbaso (1958) found spondylarthrosis in 20.6% and scoliosis of the lumbar part of the spine in 23%. The radiological findings showed a clear relationship to age and length of exposure (Tables 14 and 15).

Table 14. Pathological radiological findings in the spinal column related to age of bus driver (Barbaso 1958)

Age group (in years)	Incidence of pathological findings (%)
20–30	28
30–40	42
40–50	44
50–60	53

Table 15. Pathological radiological findings in the spinal column related to professional driving years of bus drivers (Barbaso 1958)

Professional years	Incidence of pathological findings (%)
Up to 5	26
5–10	36
10–15	38
15–20	58
More than 20	55

In a research project financed by the American National Institute for Occupational Safety and Health (NIOSH), Gruber and Ziperman (1974) conducted medical examinations on long-distance bus drivers. Using an age-correlated "health-index," it was possible to make a comparison with the general population. The authors found a significant decrease in the health index with the increase of working years with respect to skeletal and muscle changes.

In another comparative experiment, Gruber (1976) found a significantly higher incidence of premature degenerative processes in the vertebral column for long-distance truck drivers compared with bus drivers. This can be explained by the generally higher vibration stress in trucks as opposed to busses.

A research team from Gorki (1975), cited by Heide (1978), reported on experiments with 197 truck drivers and 133 taxi and bus drivers. According to this report, 41% complained of pain in the region of the lumbar part of the spine after work or after physical stress. In 28%, the diagnosis was chronic nerve root irritation in the lumbosacral region. Radiological examinations showed degenerative changes such as spondylarthrosis and spondylosis in 40% of all drivers. The percentage of radiologically proven injuries to the spinal column was statistically significant in the long-distance truck drivers (56.7%). In the taxi drivers the cervical spine suffered the most, but in the truck drivers it was the thoracic and lumbar parts of the spine. Comparison with the literature showed that degenerative changes occur three times more frequently in drivers up to the age of 30 years than in the average population.

Schoknecht and Barich (1978) carried out medical examinations on a group of 809 bus drivers and a control group of 425 office workers. They wanted to clarify the question of whether radiologically detectable changes occur more often in the vertebral column in professional drivers. An X-ray film of the lumbar part of the spine was made in two planes on all drivers examined. The radiological-epidemiological study showed no significant increase in job-related injuries to the spinal column in the drivers.

Garbe (1981) has examined the health conditions and health risks of bus drivers. The group subjected to vibration comprised 811 male bus drivers between 40 and 49 years of age, who could prove that they had had at least 5 years of activity as a bus driver and 10 years' activity in professional driving. In 39.6% of the drivers examined, pathological findings were found in the area of the lumbar part of the spine and in 10% in the area of the thoracic part of the spine.

6.4.1.3 Drivers of Earth-Moving Equipment

Kunz and Meyer (1969) analyzed complaints and defects in the vertebral column in a group of 52 drivers of heavy construction machinery after long-term exposure to whole-body vibration. In 49 workers a radiological examination of the spinal column was carried out. Of these drivers, 67% were subjectively free of disorder; 19% had given up driving or wanted to give it up because of stubborn, recurring back pain that did not respond to therapy. Of the drivers exposed to vibration, 14% indicated recurring back pain that had led to temporary inability to work, but which responded to therapy and did not lead to change of profession. About one-half of the 35 drivers who had no complaints nevertheless suffered from the following problems in the spinal column or pelvis: hip-joint diseases, slanting of pelvis after fracture of the lower leg, disc hypoplasia, pseudospondylolisthesis, spondylolysis, thoracic and lumbar Scheuermann's disease. In these cases, the author discounted any causal relationship with the driving activity. In a total of eight cases of discopathy and radiologically demonstrated narrowing of the intervertebral space, the correlation with driving activity was considered, especially since the drivers in question had been exposed to vibration for more than 10 years.

On the basis of these experimental results, the authors made the requirement that drivers of heavy construction equipment undergo an initial radiological examination to determine the condition of the spinal column before taking up their duties.

According to Cremona (1972), who examined the spinal column of workers in the iron and steel industry, as well as in ore mining, lumbar pain and low-back pain occurred in 70% of the drivers of heavy equipment exposed to whole-body vibration.

When 222 drivers of heavy trackless vehicles in a potash mine were examined (initial and repeat examinations), Franke (1978) found no changes in the condition of the spinal column in the follow-up examinations carried out after 2 or 4 years of exposure.

The American National Institute for Occupational Safety and Health (NIOSH), in an epidemiological study by Milby and Spear (1974), express a point of view with regard to the relationship between stress from whole-body vibration and the incidence of illness in drivers of heavy earth-moving equipment. The study was based on a questionnaire given to 1,865 drivers of such machines from northern California and attending physicians. For comparison, a control group was used (2,071 workers who worked outdoors but were not exposed to vibration).

The physicians' questionnaire dealt with the information requested for the diagnosis of the drivers. The drivers were confronted with the question: "In your job experience which do you feel has been the most damaging to your health: noise, heat, fumes, or dust?" Nothing was asked about vibration stress and no result could therefore be presented. Nevertheless, the length of the activity was queried. Statistical analysis indicated that there was no proof of higher morbidity (with age taken into account) in men subjected to whole-body vibration compared with the control group not exposed to vibration. It was pointed out, however, that there is a selection process where drivers with serious disorders to the vertebral column quit their jobs and thus avoid additional stress from whole-body vibration.

In the study it is also noted that only the diseases of the male sex organs (including the prostate) show a connection with whole-body vibration stress. These findings were rejected, however, by NIOSH as false statistical evidence. From a clinical point of view, this epidemiological study raises considerable doubt as to its accuracy.

In the framework of a research project, Köhne et al. (1982) have attempted to answer the question of the degree to which whole-body vibration causes permanent injuries to health. They carried out their experiments on a group of 352 drivers of heavy earth-moving equipment in the brown-coal areas of the Rhein region, as well as on a control group of 315 workers who were not exposed to vibration (surface mining).

This epidemiological study was based on a questionnaire on vibration stress in general, administered by physicians immediately after a workshift, as well as a morbidity examination. It was supplemented by the evaluation of available X-ray films of the lumbar spine of 251 drivers of earth-moving machines who had been exposed to stress from whole-body vibration for more than 10 years. These results were compared with the control group. Furthermore, the X-ray films of individual vertebral segments of 176 drivers of earth-moving machines were compared with the group that was not exposed to vibration, without taking into consideration the duration of vibration exposure.

Of all complaints, the drivers complained mostly of reduced health and well-being during and after the work shifts. An appreciably higher percentage of them also complained of spinal column disorders than did the workers not subjected to vibration stress (Fig. 46).

The disorders in the drivers of earth-moving equipment were distributed as follows: 68.7% lumbar vertebral column, 6.8% thoracic spine, and 18.2% cervical spine. This also largely corresponds to the distribution of the pathological radiological findings available for the individual segments of the vertebral column.

When 149 drivers of earth-moving equipment were questioned concerning their perception of health disturbances immediately after a work shift, back

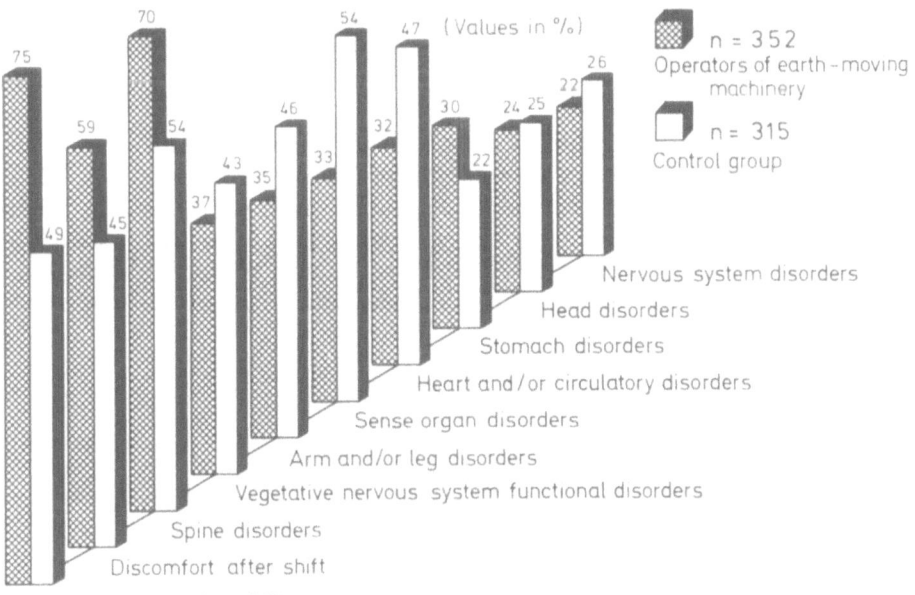

Fig. 46. Incidence of complaints of drivers of earth-moving equipment in comparison with a group not exposed to vibration (surface mine workers; Köhne et al. 1982)

pain was given most frequently, followed by fatigue and deafness. The percentage distribution of reported health disturbances is given in Fig. 47.

The health disturbances reported by the drivers of earth-moving machines immediately after the work shift showed a clear age relationship (Table 16).

In the morbidity study of 352 drivers of earth-moving equipment and control group of 315 surface-mine workers, a so-called lumbar syndrome was determined in 81% of the drivers exposed to vibration. In the workers of the control group, the percentage was only 53%. The diagnosis "lumbar syndrome" represents a clinical unity. The clinical picture, also known as lumbar spine syndrome, is composed, according to Krämer (1978), of diseases of the spinal column that are directly or indirectly caused by degenerative processes in the lumbar vertebrae. Included here especially are spondylosis of the lumbar part of the spine (disk degeneration with reactive bone spurs on the vertebral borders), spondylarthrosis of the lumbar spine (degenerative changes in the vertebral joints, mostly disk degeneration), and spondylosteochondrosis of the lumbar spine (disk degeneration, with separation of the body and end plates of the vertebral bodies). These processes were only included if radiologically proven changes had occurred in the spinal column with disk-related complaints (pain or disturbances in function, arising from the lumbar area). Clinical pictures, such as sciatica (lumbar syndrome with involvement of the sciatic nerve) and lumbago (acute form of the lumbar syndrome), as well as spondylolisthesis, were also included in the lumbar syndrome category.

Figure 48 shows the percentage of illnesses and disturbances to health (morbidity study) in the drivers of earth-moving machines in comparison to the surface-mine workers who were not exposed to vibration.

In three cases, the diagnosis "fracture of the spinal process of a vertebral body in the cervical part of the spine" was established in the drivers of earth-moving equipment. The diagnosis of injury to the cervical vertebral column was confirmed by radiology (Fig. 49).

The distribution of morphological changes of the spinal column is presented in Table 17 with regard to particular segments. When discussing this table, it should be remembered that individual diagnoses, which had partly been made by other physicians, clinics, etc., that were not a part of the study were not changed as regards terminology. Furthermore, it should also be taken into consideration that in diseases of parts of the spinal column, it is possible to use a

Table 16. Complaints (back pain) related to age of drivers of earth-moving equipment (Köhne et al. 1982)

Age group (years)	Incidence of complaints (in %)
20–29	35
30–39	28
40–49	54
50–59	67
Total	45

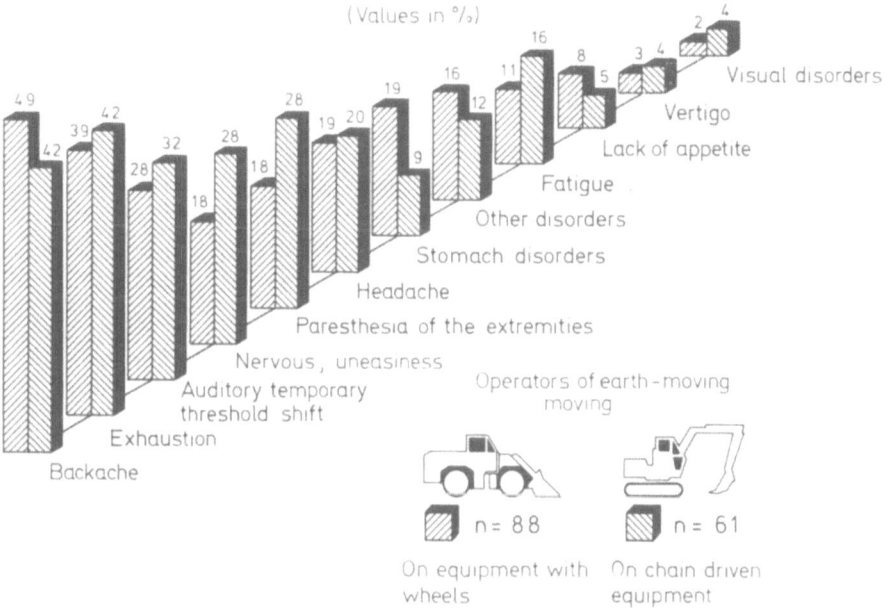

Fig. 47. Incidence of health disturbances reported by drivers of earth-moving machinery immediately after 8-h stress from whole-body vibration (Köhne et al. 1982)

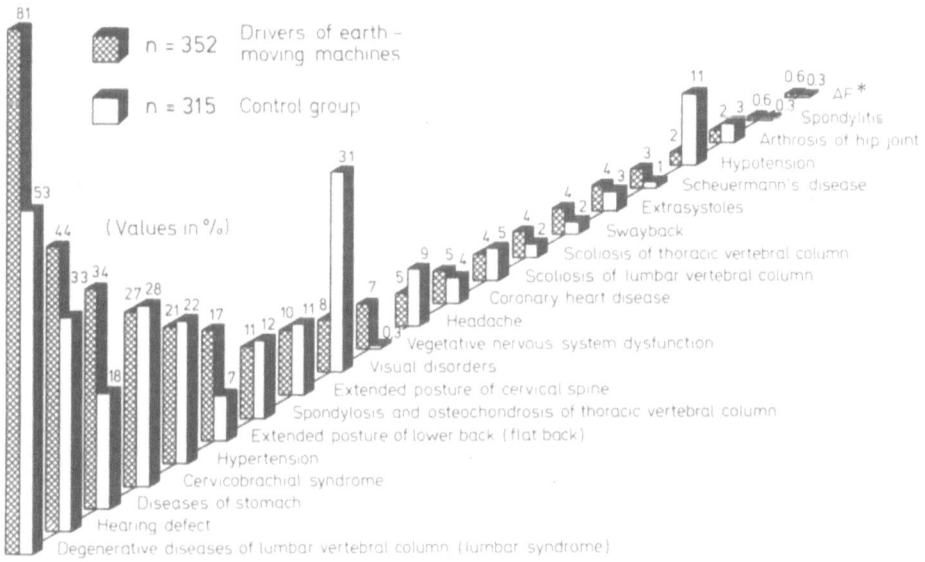

*AF, avulsion fracture of a spinous process of the lower cervical spine

Fig. 48. Incidence of various diagnoses in drivers of earth-moving machinery who were exposed to vibration (n = 352) in comparison with surface mine workers not exposed to vibration (n = 315) (Köhne et al. 1982)

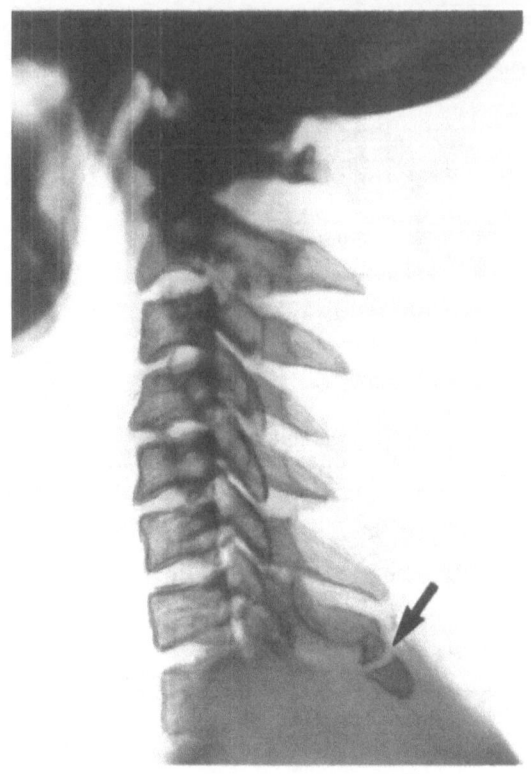

Fig. 49. Fracture of the spinal process of the C7 vertebral body in a driver of earth-moving equipment who had been subjected to whole-body vibration (Köhne et al. 1982)

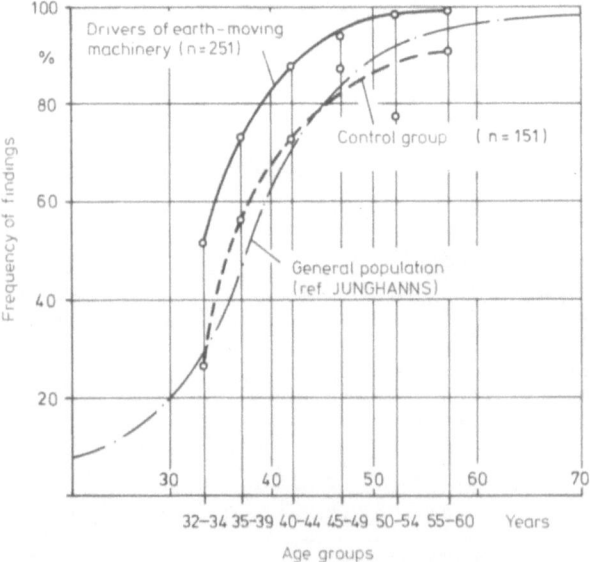

Fig. 50. Radiologically proven morphological changes in the lumbar spine (degenerative diseases) of drivers of earth-moving equipment, with at least 10 years of exposure to whole-body vibration compared with a control group not exposed to vibration (distributed by age groups). Also shown are the "morphological findings" in the thoracic and lumbar spine in the average population according to Junghanns (1931), Köhne et al. (1982) and Junghanns (1979)

Table 17. Incidence of morphological changes in areas of the spine in drivers of earth-moving equipment in comparison with a control group (Köhne et al. 1982)

Spinal column area and diagnosis	Drivers of earth-moving equipment ($n=352$)		Control group not subjected to vibration ($n=315$)	
	Absolute	Relative %	Absolute	Relative %
Cervical spine				
Spondylosis deformans	40	11.4	34	10.7
Spondylosteochondrosis	29	8.2	23	7.3
Cervical syndrome	22	6.3	22	7.0
Uncovertebral arthrosis	4	1.1	8	2.5
Thoracic spine				
Spondylosis deformans	24	6.8	22	7.0
Spondylosteochondrosis	14	4.0	11	3.5
Thoracic syndrome	1	0.3	3	1.0
Scheuermann's disease	9	2.6	3	1.0
Kyphosis	10	2.8	0	0
Lumbar spine				
Spondylosis deformans	166	47.2	57	18.1
Spondylosteochondrosis	67	19.0	36	11.4
Lumbar spine syndrome	40	11.4	66	20.9
Damage to intervertebral discs	11	3.1	16	5.1
Spondylolisthesis	8	2.3	7	2.2
Pseudospondylolisthesis	2	0.6	0	0
Scoliosis	14	4.0	17	5.4
Extended posture (flat back)	46	13.1	15	4.8

manifold variety of terms: e.g., spondylosis and extended posture of the spine. The lumbar part of the spine was involved in the majority of pathological processes. For 251 of the 351 drivers of earth-moving equipment who had been exposed to whole-body vibration for more than 10 years, X-ray films of the lumbar part of the spine were available for use. The distribution of incidence of radiologically proven changes in the various age groups of this group showed that, in comparison with a group of workers who had not been exposed to vibration stress, there was an earlier and markedly higher incidence of morphological changes in the region of the lumbar spinal column.

The significance of the results presented in Fig. 50 was proven statistically at the 0.1% level. Drivers of earth-moving equipment who had been exposed to at least 10-year stress from whole-body vibration showed, therefore, a much higher incidence of positive, radiologically proven, pathological findings, in the sense of premature wear and tear (degenerative process) of the lumbar spine, than those workers not exposed to vibration.

In summary, the study by Köhne et al. (1982) indicates that morphological changes of the spinal column, particularly in the lumbar region, are manifested more frequently and more prematurely in drivers of earth-moving equipment and that these changes must have been caused by stress from whole-body vibration. Likewise, the fracture of the spinal process in the lower region of the cervical spine has to be considered as vibration-induced damage to the spinal column. The subjective complaints of drivers exposed to vibration stress and the increase in morbidity regarding the so-called lumbar syndrome confirm the objective radiological findings in the spinal column, particularly in the lumbar region.

6.4.1.4. Railroad, Ship and Airplane Crews, and Employees in the Concrete Industry

During radiological examinations of the cervical and lumbar regions of the spine in 76 railroad employees, Arnautova-Bulat (1979) found signs of spondylosis and spondylarthrosis more frequently than in a control group of workers. Spondylosis in the cervical region was found more frequently in younger railroad employees than in older fellow workers. The vertical and horizontal vibration encountered during travel, which depended on the condition of the tracks and the speed of the train, was assumed to be the decisive factor in damage to the spinal column.

Louyot et al. (1954) carried out radiological examinations of the spinal column in 156 locomotive firemen of the French national railway, the SNCF; 90 firemen were examined over a period of 5 years and the remaining 66 in the course of 1 year. In 80% they found radiologically confirmed changes in the spinal column (partial fractures in the vertebral bodies, tori on the anterior border of the vertebral bodies, and progressive wedgelike flattening of the vertebral bodies in the dorsolumbar transition).

Although 30 of 78 X-ray films showed "fatigue" of the intervertebral disk, there were only two cases of hip-joint symptoms. This proved to the authors that the changes observed in the spinal column were not caused by lifting heavy loads. The authors did not discuss the effect of mechanical vibration as a potential cause of the processes observed.

Kersten (1966) researched the connection between diseases of deep-sea fishermen and the different kinds of work they perform; he discovered that diseases of the support apparatus and locmotion system are clearly related to work stress. The work conditions in deep-sea fishing are characterized by, among other things, vibration derived from the propulsion system of the ship.

From 1960 to 1964, according to the results of the authors, diseases of the support and locomotion apparatus leading to unfitness for work took fifth place in the morbidity of the fleet. In 1964, 10.9% of all sailors were unfit for work because of diseases of the support apparatus and locomotion system.

X-ray films of the vertebral column were taken of 220 deep-sea fishermen who were free of complaint when examined for fitness for sea duty. Of the sailors examined, 87% showed the following radiologically determined changes in the vertebral column: extended posture (flat back) of single vertebral segments, narrowing of the vertebral disks and marginal spurs of marginal projections (spondylosis). These changes were localized mainly in the area of the lower thoracic spine. Changes in the cervical and the lumbar regions did not occur as frequently as in the lower thoracic region.

In the opinion of the authors, the slow starting and chronically progressive injuries due to overstrain (especially in the spinal column) were unavoidable in sailors employed in deep-sea fishing at that time. However, in this professional group, the proportion of stress caused by physical hard work, cold, and vibration still remains to be clarified.

Regarding strain in aircraft, Beck (1973) considered certain injuries to health and complaints, particularly in the spinal column. He held that the vibration, shock and shaking occurring in aircraft (to a great degree in helicopters) is responsible for injuries to the spinal column in pilots. The relatively numerous problems in the spine expressed by helicopter pilots are well known and have been observed in air forces of many countries.

Fischer et al. (1980) discovered vibration-induced injuries to the spinal column in helicopter pilots by means of comparing the findings of clinical and radiological examinations of 221 helicopter and 226 jet pilots. Of the helicopter pilots examined both clinically and radiologically, 76% complained of pain in the spine. Of these complaints, 82.7% were considered by the pilots to be in the lumbar area and 15.5% in the cervical part of the spine. The helicopter pilots showed significantly more problems in the spinal column than the jet pilots, above all in the region of the lumbar part of the spine. This could also be attributed to the frequently unfavorable sitting position.

In the helicopter pilots with more than 3,000 flight hours, the percentage of radiologically confirmed changes in the spinal column was 80%. In contrast to the jet pilots, the helicopter pilots were found to have a high incidence of spondylosis and osteochondrosis.

Beck (1981) qualified the data on the radiologically confirmed degeneration of the vertebral column in helicopter pilots. The changes in the spinal column, encountered more frequently in helicopter pilots than in jet pilots, particularly in the lumbar sacral region, indicate no more degeneration than in nonflyers. It is assumed that the differential selection procedures in helicopter pilots cause increased degeneration of the spinal column. It must still be taken into consideration, however, that helicopter pilots, as opposed to "nonflyers," have stricter requirements regarding fitness.

Rumyantsew and Chumak (1966) reported on bone changes in the spinal column in 78 concrete workers after exposure to whole-body vibration. They found changes such as: spondylosis deformans, intervertebral osteochondrosis, and calcification of the intervertebral disks. The lumbar region of the spine was examined radiologically in workers who had been exposed to high-frequency whole-body vibration (50 Hz). Degenerative processes, like spondylosis deformans, were found in 50% of the workers who ranged in age from 24−36 years. Osteochondrosis was found in 14.8% and calcification of the intervertebral disks in 22.2%. In a control group of 52 persons 25−40 years old, only 7.7% showed spondylosis deformans. There was no information on what vibration stresses and physical stresses the concrete workers had been subjected to, and therefore no evaluation could be carried out.

6.4.1.5 Prevalence of Pathological Changes in the Spinal Column (Summary)

The results of the aforementioned studies are summarized (Tables 18−21) for various occupational groups with regard to the established incidence of radiologically supported pathological findings in the vertebral volumn. Unfortunately, most of these publications contain no data on the vibration stress that the examined group had been exposed to. Heide (1978) found evaluative measurement data on vibration exposure in 37% of the 46 publications she reviewed. In recent years, there have been numerous experiments with results which − with certain reservations − can be used to estimate the amount of stress in different vehicles and work machines (see section "Work-Related Stress from Whole-Body Vibration," p. 9).

In many experiments there was no control group, and only in exceptional cases could the general health of the spinal column be followed in long-term studies from the beginning of the occupation over a few years or longer (e.g., Dupuis and Christ 1972; Franke 1978). In spite of this fact, the study material shows a tendency that vibration stress occurring in drivers of earth-moving machines and at one time in tractor drivers (today vibration stress on tractors has

Table 18. Incidence of pathological radiological changes in the spinal column in tractor drivers

Authors	Number of cases examined	Incidence of pathological findings[a] (in %)
Fishbein and Salter (1950)	378 (physicians queried)	None
Rosegger and Rosegger (1960); Rosegger (1966, 1970)	310	71
Lavault (1962)	107	33
Zimmermann (1964)	137	80
Christ (1963) First experiment	211	50
Dupuis and Christ (1966 b) 1. Follow-up experiment	137	69
Dupuis and Christ (1972) 2. Follow-up experiment	106	80
Kubik (1966)	400	68
Seidel and Tröster (1970)	60 (Questionnaire)	None
Köhl (1975)	49	None "Higher rate of findings in the spinal column"
Schulze and Polster (1979)	103	80

[a] Percentages rounded off after the decimal point

Table 19. Incidence of pathological radiological changes in the spinal column in drivers of utility vehicles (trucks, pick-ups, buses)

Professional·group Author	Number of cases examined	Incidence of pathological findings[a] (in %)
– Drivers of heavy-duty trucks Schmidt (1969)	117	79
– Drivers of pick-up trucks Kristen et al. (1981)	94	83
– Bus drivers		
Barbaso (1958)	213	44
Collection of authors Gorki (1975), cited by Heide (1978)	330	57
Schoknecht and Barich (1978)	809	None
Garbe (1981)	811	40 (lumbar spine) 10 (thoracic spine)

[a] Percentages rounded off after the decimal point

Table 20. Incidence of pathological radiological changes in the spinal column in drivers of earth-moving equipment

Author	Number of cases examined	Incidence of pathological findings[a] (in %)
Kunz and Meyer (1969)	49	50
Köhne et al. (1982)	352	81

[a] Percentages rounded off after the decimal point

Table 21. Incidence of pathological radiological changes in the vertebral column in railroad personnel, ship and airplane crews, and employees in the concrete industry

Professional group Author	Number of cases examined	Incidence of pathological findings[a] (in %)
Ship crews Kersten (1966)	220	87
Air crews Fischer et al. (1980)	136 (helicopter pilots)	80
	143 (jet pilots)	47
Railroad personnel Louyot et al. (1954)	78 (firemen)	80
Arnautova-Bulat (1979)	76 (railroad workers)	Up to 62
Concrete workers Rumantsew and Chumak (1966)	78	87

[a] Percentages rounded off after the decimal point

been reduced by means of improvements in seat suspension) may lead to an increased incidence of pathological findings in the vertebral column.

6.4.1.6 An Occupational Medicine Appraisal of Changes in the Spinal Column Arising from Vibration Exposure

Heide (1978) believes that pathological changes in the cervical area of the spine, in persons exposed to vibration stress, are caused less by vibration than by the result of static strain and the necessity to maintain a certain posture. In this regard, Rosegger and Rosegger (1960) and Köhne et al. (1982) have also observed cases of fractures of the spinal process in the cervical region, in which faulty posture, static muscular stress of the neck, and long-term vibration stress have probably led to fatigue fracture.

On the other hand, even though the literature shows that there is increased susceptibility to stress in the lumbar region from whole-body vibration, this cannot be assessed as a specific sign of strain. Current opinion does not hold that there is a specific vibration symptomatology or vibration disease, such as that described by Andreeva-Galanina (1957, 1961), among others.

The degenerative processes observed in the spinal column apparently deal with premature and developing wear of the skeletal system, which is the result of long-term exposure to vibration stress. This has been shown by numerous experiments, epidemiological studies, and X-ray analysis.

The formative mechanism of this premature degenerative disease of the spinal column encountered has so far not yet been clarified. Instead, a series of hypotheses have been formulated for the formative mechanism of which, above all, two causes are suggested (Dupuis 1980 b):
1. Mechanically caused overstrain
2. Disturbances in the metabolic processes of the intervertebral disks

The hypothesis of mechanically caused overstrain assumes that vibration exposure of high intensity and duration during sitting produces unphysiological stress. Such stress does not occur under the natural living conditions to which man has been adapted for thousands of years. Included in these are the occurrence of resonance of the vertebral column, as well as random, i.e., irregular vibration exposure. Various authors using different methods have shown experimentally that resonance vibration of the spinal column occurs between 3 and 6 Hz, with a maximum at 4−5 Hz in the complex mass-spring-damper system of the trunk when sitting. This is exactly the frequency range in which many vehicles and work machines operate. Resonance applies to biological as well as inanimate materials and leads to increased tissue strain. The constant compression and stretching of the spinal column may fatigue tissue (regarding its function) and reduce its ability to withstand stress if there is insufficient opportunity to recover. In this respect, Junghanns (1979) points out that intervertebral disks, as well as bones and joints, react only after long-term exposure to mechanical stress and occupationally related vibration stress. The compression forces are probably especially harmful if the spinal column is at the extreme of its movement. For example, Rosegger (1966, 1970), Junghanns (1979), and Dupuis (1980'b) have expressed opinions of this sort.

Body posture is also considered to be an important influence, according to Dupuis (1969). Extending the spinal column with a flat back increases the risk of forming degenerative symptoms in the spinal column during whole-body vibration, because the vibration transmission is increased.

When looking into the mechanics of the spinal column, the nerves also need to be taken into consideration (Junghanns 1979). Thus, the close interconnection of the spinal cord and the segmental nerve roots can be the reason why impairment of the nerves is caused by mechanical vibration. Whereas in the usual movement and stress of the spinal column there is no danger, deviations of the form and posture of the spinal column and pathological changes in the region of a motion segment can change this situation for the worse.

Another hypothesis regarding the formation of vibration-caused injuries to the spinal column maintains that the intervertebral disks, as bloodless tissue, are dependent on the diffusion of neighboring tissues. According to Junghanns (1979), vibration may affect normal metabolic exchange in a disadvantageous manner because the fluids, at the junction of the vertebral body, and the disks are only moved back and forth as a result of vibration. If diffusion disturbances occur, the consequences could be accelerated degeneration of the vertebral disks. In the opinion of Junghanns (1979), this could be possible when certain frequencies are encountered that, similar to the vibration-caused vasospastic syndrome (vibration white fingers), could lead to disturbances in the capillary blood flow.

In addition to the formative mechanism, the question of which individual and exogeneous conditions also contribute to the acceleration and intensification of degenerative processes, when under vibration stress, is also important.

Included in the individual factors are: the age of the person exposed, the age at the beginning of the occupation, the endogenous conditions of the spinal column, and the living habits of the person involved (Dupuis 1980 b). With increasing age, the amount and composition of the bone substance change, as does the water content of the intervertebral disks. Therefore, the decreasing stability and elasticity of the spinal column as a whole must be taken into consideration. However, this indicates a decrease in the ability of the spinal column to withstand mechanical vibration.

As long as the growth of the spinal column is not yet complete (growth ends at 20−23 years of age), the age at the beginning of exposure to vibration in the occupation can be of importance. Although the reaction of growing bones to vibration has so far not been sufficiently researched, animal experiments on pregnant mice exposed to vibration have proven that there is a decrease in growth time in the embryonal spinal column through reduction of cell division (Bantle 1971). Apparently, disturbances in the growth zones of the undeveloped spinal column take place, but it is not yet certain whether a direct vibration effect or a disturbance in the internal secretory hormonal control of growth is responsible here. Thus, it appears certain (Rosegger 1970) that the young, incompletely developed spinal column is less capable of resisting the mechanical vibration encountered than the spinal column of a tractor driver. Youth subjected to stress can therefore be regarded as particularly endangered, whereas after completion of the 25th year of age, a special disposition for a healthy spinal column can no longer be expected (Heide 1978).

There are no clear data in the specialist literature regarding the effect of mechanical vibration on persons with varying constitution and predisposition. According to Raymond (1956), people with leptosomatic constitutions are more sensitive. However, in analogy to the positive correlation between body weight and osteochondrosis and spondylosis deformans (Häublein 1977), one can assume that obesity represents a predispositional factor for vibration-caused degenerative illnesses (Heide 1978). On the other hand, one must conclude that well-developed back, thoracic, and abdominal musculature can represent a positive function for the support of the spinal column during vibration stress, so that body weight alone is not a predispositional factor.

Undoubtedly, the endogenous condition of the vertebral column can have a meaningful influence on resiliance. Although here there also appears to be no definite knowledge, one can probably assume that an unhealthy spinal column is more prone to overstrain from mechanical vibration than a healthy spine. Scheuermann's disease, with the endogenous weaknesses of the upper plate of the vertebrae, may have an unfavorable influence on the resiliance of the spinal column (Dupuis and Christ 1972). This also applies to scoliosis of medium and severe degree, spondylolysis and spondylolisthesis, as well as severe forms of degeneration (osteochondrosis intervertebralis, spondylosis deformans and arthrosis deformans).

Furthermore, individual living habits must also be taken into consideration with regard to strain on the spinal column. It has already been pointed out that well-developed musculature of the trunk, above all the back muscles, supports the spinal column better and therefore can also reduce the shock-induced twisting and bending of the spinal column. In contrast, pronounced muscle insufficiency reduces the resiliency of the spinal column. As every muscle can be trained, the trunk muscles should be trained more than previously in the form of reasonable, regular sport activity, as a prophylactic means of improving endogenous conditions.

Concerning the exogenous stress factor, the "dose" of vibration stress is the most important, i.e., the intensity and duration of exposure encountered on a daily and yearly basis, as well as over the entire occupational life. Including age-conditioned changes, many experiments have thus pointed out that an increase in prevalence or worsening of pathological changes in the spinal column are dependent on the length of exposure (Louyot et al. 1954; Barbaso 1958; Rosegger and Rosegger 1960; Christ and Dupuis 1966b; Gruber and Ziperman 1974; Köhl 1975). So far it has not yet been possible to estimate the definite minimum exposure time before degenerative diseases of the spinal column begin to occur with a higher prevalence.

Evaluation of epidemiological findings in groups of people subjected to especially strong vibrations is particularly difficult when there is a fluctuation caused by people who change their occupation because they feel the stress to be painful and unbearable. This results in a selection of persons who are more capable of resisting the stress, as can be observed, for example, in drivers of earth-moving equipment (Milby and Spear 1974; Köhne et al. 1982).

Regarding the reasonable daily duration of exposure without risk to health, some indications are given in the form of standard curves in the German guideline VDI 2057 and the international standard ISO 2631 (see Fig. 54). According to the present state of knowledge, from the point of view of the intensity of vibration, these guidelines appear to be a valid foundation for the evaluation of the question as to whether a certain stress represents a risk to health of the vertebral column or not. So far, current scientific knowledge does not yet permit the representation of a quantitative dose-effect relationship. The standard curves, namely VDI 2057 and ISO 2631, however, make use of the current state of knowledge and a region (dependent on the length of exposure and the intensity of vibration) in which a risk of degenerative changes in the spinal column must be reckoned with during long-term exposure. This point of view is also taken by Heide (1978) and Heide and Seidel (1978) on the basis of their comprehensive evaluation of 29 epidemiological experiments, in which some 24,000 persons were exposed to whole-body vibration. Furthermore, the Highway Safety Research Institute in Michigan has recently established the state of knowledge on the relationship between truck vibration stress and pathological effects. Segel et al. (1981) in this way have come to the conclusion that mechanical vibration in such vehicles is definitely not the primary reason for degenerative processes in the spinal column but that, however, the occupational group of truck drivers is at greater risk of premature and more severe development of these diseases. In 1980, experts in the International Labor Office (ILO) recognized, in principle (see section "Work-Related Diseases in National and International Regulations," p. 88), the occupational origin of diseases of the spinal column as a result of vibration in the whole body, especially in drivers of power-driven machines and tractors. However, the exact risk, as expressed as the incidence of these diseases in relation to exposure data, for example, is not yet known.

In individual cases it is difficult to prove that degenerative changes in the vertebral column are causally connected to stress from whole-body vibration, because comparable changes in the vertebral column can be expected in a large portion of the total population in the sense of normal wear and tear. With age, differentiation of such changes is made even more difficult. However, when degenerative changes with pronounced symptomatology occur in the vertebral column of young people who have been exposed to strong stress from whole-body vibration (according to VDI 2057 or ISO 2631), then they are judged to be most likely to be work-conditioned if the changes can be regarded as premature from the point of view of age. The longer the exposure to vibration has been, the more probable it is that a connection can be traced to stress from whole-body vibration.

6.4.2 Digestive System Diseases

In addition to spinal problems, persons who have been exposed to mechanical whole-body vibration over a long period are often observed to have complaints and illnesses centered in the stomach and digestive system generally. As shown in the section "Vibration Behavior of the Internal Organs" (see p. 39), X-ray cineradiographic examinations have shown that when man is under the influence of sinusoidal or random vibration, resonance movement of the stomach results, with a maximum between 4 and 5 Hz. The tissues in this region are then under especially intensive strain, so that the vibration stress must be regarded as an important etiological factor in the formation of stomach complaints and illnesses. Again, certain occupational groups are affected.

6.4.2.1 Tractor Drivers

From the point of view of complaints and illnesses of the digestive system, many studies have been conducted on tractor drivers in particular. Paulson (1946), cited by Heide (1978), was the first to report on the high incidence of stomach illnesses during and after stress from whole-body vibration. Rentsch (1960) also demonstrated higher morbidity on the basis of clinical and radiological examinations and on a questionnaire given to 915 tractor drivers. The number of tractor drivers with stomach problems lay just barely over the average of his control group (agricultural workers, excluding tractor drivers). For stomach and duodenal ulcers, the author found a morbidity of 4.9% in the tractor drivers.

Rosegger and Rosegger (1960) carried out the first large-scale epidemiological study. On the basis of observation of a relatively high morbidity regarding stomach diseases in tractor drivers, they conducted radiological examinations of the stomach (stomach-intestinal passage) in a total of 322 tractor drivers. Of these drivers, 51.2% complained of stomach problems; in 89.7% of these vibration-stressed workers, the stomach complaints mentioned first occurred during their activity as tractor drivers. The radiographs showed pathological findings in the stomach of 76.1%. The authors evaluated 74.3% as "acute" and 25.7% as chronic. Among these so-called acute findings, approximately two-thirds had dropped stomach (gastroenteroptosis). The remaining 37.9% "acute" findings showed other pathological findings, such as hypersecretion, gastritis, hypersecretory gastritis, and combinations with a dropped stomach.

Among other deformations included in "chronic" findings, the authors included bulbus duodeni (the condition after the course of a duodenal ulcer is over), various changes and combinations between "acute" and "chronic." Dropped stomach also occurred in a third of those examined (35.1%). The authors established that the number of stomach complaints and pathological radiological findings in the stomach of the tractor drivers climbed with the years of work. After more than 4 years of activity, the number of chronic stomach diseases also increased steadily. According to the study, tractor drivers with a leptosomatic constitution are more strongly at risk from the point of view of formation of stomach diseases. Since the corresponding epidemiological data on the control group lay clearly below the morbidity of those exposed to vibration, the authors were almost completely certain that the equipment itself, i.e., the tractor without a suspension system, is to be regarded as etiologically responsible, in addition to the work and life styles of professional tractor drivers.

As regards the question of the increased incidence of stomach ailments in tractor drivers, the experiments conducted and the social-hygienic views taken have been reviewed in the dissertation by Heppner (1961), cited by Heide (1978). A total of 196 tractor drivers were examined both clinically and radiologically, with emphasis on their stomach complaints. A very high percentage of stomach problems was found in the drivers who had been subjected to whole-body vibration: after 2 years in this occupation, 42.1% had stomach complaints; after 4 years 68.0%; after more than 6 years' exposure, 80.7%.

In a study of the relationships between work or work environment and the health of tractor drivers, Kubik (1966) found gastrointestinal complaints and serious disturbances in the stomach-intestinal region in 20% of 400 tractor drivers. Among other symptoms, these problems manifested themselves through loss of weight, chronic lack of appetite, pain in the stomach area, and functional and anatomic changes in the stomach. The author blamed mechanical vibration as the most damaging factor in the development of these disturbances to health, as the majority exhibiting these problems worked on tractors, which have a high vibration level. However, he observed no more cases of gastroenteroptosis than is seen in the rest of the population.

In a study attempting to isolate potential spinal problems in 137 young tractor drivers, Dupuis and Christ (1966b) and Christ and Dupuis (1968) have established a relationship between the extent of stomach complaints in people exposed to vibration and the amount of time they drove a tractor (Table 22). It should be noted that the highest age of the population examined was only 23 years.

Seidel and Tröster (1970) also found stomach ailments in a group of 60 tractor drivers. This type of ailment took third place of all the complaints encountered in the epidemiological study. Approximately 17% of the drivers examined complained of stomach pain; 70% of the drivers with stomach problems attributed them to the use of their machines.

Table 22. Incidence of stomach problems related to annual exposure to vibration (Christ and Dupuis 1968)

	Persons with < 700 tractor-driving h/year	Persons with 700–1200 tractor-driving h/year	Persons with > 1200 tractor-driving h/year
Total number of persons	$n = 44$	$n = 77$	$n = 16$
Incidence of stomach problems	18.2%	23.4%	25.0%

In a large epidemiological experiment, Köhl (1975) reported on the dangers to health in tractor drivers. The study, which is based on a questionnaire administered to 536 tractor drivers in the Swiss canton of Vaud, disclosed a 24% morbidity as regards disturbances of the intestinal tract. This is statistically highly significant. According to the results of the author, the stomach complaints occur more frequently at the beginning of occupational activity if the driver is young. ·

6.4.2.2 Truck and Bus Drivers

Raymond (1956) has investigated the morbidity of stomach complaints in a large epidemiological study of the effect of mechanical vibration on drivers of heavy trucks. In the framework of clinical and radiological examinations on a group of 311 drivers, he found that this group is at higher risk of having functional gastrointestinal problems without organic damage. However, the results were not statistically evaluated.

While conducting clinical examinations on drivers of heavy-duty trucks and office employees in order to assess the question of occupationally caused degeneration injuries to the spinal column and to the joints of the upper extremities, Schmidt (1969) also asked the 117 drivers if they had stomach problems. Of these heavy-duty truck drivers, 17% admitted that they suffered from stomach problems either occasionally or often. An ulcer was confirmed by radiology in 9.3% of these drivers.

Gruber and Zipermann (1974) reported on the relationship between stress from whole-body vibration and morbidity in truck, bus, and taxi drivers. They found that there was an increased morbidity as regards illnesses of the intestinal system (peptic ulcer, appendicitis, diverticulitis, gastroenteritis, nervous stomach), which increased with exposure time. Of the 1,448 drivers examined, 16% proved to have these complaints after 10 years' exposure, 35% after 20 years, and 47% after 30 years.

A group of authors from Gorki (1975), cited by Heide (1978), demonstrated an increase in morbidity of stomach illnesses or complaints when conducting physical examinations of 330 drivers. Stomach or duodenal ulcers were found in 15.7% of the drivers. In 14.5% of the truck drivers, chronic gastritis was found, in bus and taxi drivers 12%, and in taxi drivers alone 16.7%. An increase in morbidity of chronic gastritis was found in taxi drivers.

In the framework of a morbidity study on workers under whole-body vibration stress, Spear and Keller (1976) evaluated 1,700 documents from health insurance companies. They found that persons exposed to vibration are at higher risk of illnesses of the esophagus, the stomach, and the duodenum.

Garbe (1981) has assessed the health status and health risks of bus drivers in West Berlin. The epidemiological study included 811 bus drivers between 40 and 49 years of age who were shown to have had at least 5 years' duty as bus drivers and 10 years' activity as professional drivers. When questioned, 41.3% of the drivers indicated pain and a sensation of fullness in the stomach; this percentage was greater than for the control group. When drivers were certified to be unfit for driving before they had reached the age of retirement, illnesses of the stomach and intestinal tract were counted as reasons for invalidity, a category that was indeed the fourth-most frequent illness.

6.4.2.3 Drivers of Earth-Moving Equipment

In a Swiss study by Kunz and Meyer (1969), complaints of the upper abdomen, such as dropped stomach among others, were observed in 52 drivers of heavy construction machines. These complaints often played a large role in changes in profession. Activity as a driver must frequently be discontinued as a result of stomach problems.

In an extensive research project, Köhne et al. (1982) pursued the question of a causal relationship between stress from whole-body vibration and effects on health. In order to assess the strain, 352 drivers of earth-moving equipment, who had been exposed to the effects of long-term vibration, were questioned by the company physician. For comparison, a group of probands was used of the same age group, but who were not exposed to vibration. Of the drivers of earth-moving equipment, 30% indicated stomach problems (22% in the control group).

Immediately after an 8-h shift, 149 drivers of earth-moving vehicles were also questioned regarding disturbances to health. Acute stomach problems were the complaints of 14% of those exposed to vibration stress. In order to differentiate between the ailments given, according to the type of earth-moving equipment (those with wheels and those with tracks), it was shown that the drivers of wheeled vehicles had a distinctly higher percentage of stomach problems (19%) than did drivers of tracked vehicles (9%). The stomach ailments proved to be age dependent: with increasing age the morbidity climbed.

In the above epidemiological study, illnesses of the stomach made up 34% of all illnesses and took third place below the lumbar syndrome (81%) and hearing difficulties due to noise (44%). The control group that had not been exposed to vibration showed an 18% morbidity of stomach illnesses. The kinds of illness and the incidence of each are given in Table 23 for both the drivers of earth-moving equipment and the workers not exposed to vibration (control group).

Table 23. Diagnoses of stomach problems in drivers of earth-moving equipment compared with a control group (Köhne et al. 1982)

Diagnosis	Drivers ($n = 352$)	Control group (not exposed to vibration; $n = 315$)
Duodenal ulcer	5.1%	4.4%
Duodenitis	0.6%	0.0%
Deformity of the duodenal bulb	2.0%	0.3%
Waterfall stomach	2.8%	2.0%
Gastric ulcer	8.0%	5.7%
Gastritis	8.2%	2.2%
Portion of stomach resected because of a gastric or duodenal ulcer	2.3%	2.0%

Table 24. Incidence of pathological findings and complaints in the area of the stomach in tractor drivers

Author	Number of cases examined	Diagnosis	Incidence of pathological findings[a]
Rentsch (1960)	915	Stomach and duodenal ulcers	5
Rosegger and Rosegger (1960)	322	Radiological findings (total)	76
		Gastroptosis	35
		Stomach complaints	51
Heppner (1961) according to Heide (1978)	196	Stomach complaints	
		– after 2 years' experience	42
		– after 4 years' experience	68
		– after 6 years' experience	81
Kubik (1966)	400	Stomach pain and functional and anatomical changes	10
		Gastrointestinal complaints	20
Dupuis and Christ (1966 b)	137	Stomach complaints	22
Seidel and Tröster (1970)	60	Stomach complaints	17
Köhl (1975)	536	Disturbances in region of digestive tract	24

[a] Percentages rounded off after the decimal point

Table 25. Incidence of pathological findings and complaints in the area of stomach in drivers of utility vehicles (trucks, buses)

Author	Number of cases examined	Diagnosis	Incidence of pathological findings[a] (in %)
Raymond (1956)	311	Functional and gastro-intestinal complaints	No response
Schmidt (1969)	117	Stomach complaints	17
		Stomach and duodenal ulcers	9
Gruber and Ziperman (1974)	1,448	Stomach and duodenal ulcers, nervous stomach	
		– after 10 years' experience	16
		– after 20 years' experience	35
		– after 30 years' experience	47
Collection of authors Gorki (1975), cited by Heide (1978)	330	Stomach and duodenal ulcers	16
		Chronic gastritis	12
Garbe (1981)	811	Stomach complaints (pain, bloating)	41

[a] Percentages rounded off after the decimal point

Table 26. Incidence of pathological findings and complaints in the area of the stomach in drivers of earth-moving equipment and employees in the concrete industry

Author	Number of cases examined	Diagnosis	Incidence of pathological findings[a] (in %)
Drivers of earth-moving equipment			
Kunz and Meyer (1969)	52	Stomach diseases	No response
Köhne et al. (1982)	352	Stomach complaints	30
		Various diseases of the stomach	34
Workers in the concrete industry			
Gracianskaja (1962)	145	Stomach pain	12
		Stomach and duodenal ulcers	5
		Chronic gastritis	9

[a] Percentages rounded off after the decimal point

6.4.2.4 Employees in the Concrete Industry

In clinical examinations of 145 hospitalized workers of a concrete factory who had been exposed to stress from whole-body vibration on vibration platforms for solidifying concrete, Grascianskaja (1962), cited by Heide (1977), reported that 11.8% of the cases had pain in the epigastrium (stomach pain). In addition, 5% of the workers suffered from stomach or duodenal ulcers, and 9% had chronic gastritis. The stomach ailments occurred significantly more frequently in the workers exposed to vibration than in the control group.

6.4.2.5 Prevalence of Pathological Findings in the Digestive System (Summary)

The results of the aforementioned experiments are summarized in Tables 24–26 for the various occupational groups with regard to the incidence of pathological findings and complaints established.

6.4.2.6 An Occupational Medicine Appraisal of Digestive Disorders Arising from Vibration Exposure

From the results of epidemiological studies, as well as clinical observations or experience in occupational medicine, it can be assumed that, in workers under occupational stress from whole-body vibration, there will be a statistically significant increase in morbidity as regards functional and organic stomach illnesses, in comparison with other occupational groups without vibration stress. This increase in morbidity is apparently more pronounced with time.

Etiologically, however, it cannot be concluded that whole-body vibration can be isolated as the sole triggering factor for these stomach complaints. Paulsen (1946), cited by Heide (1978), has already shown that the etiology of stomach diseases and complaints cannot be explained by one single factor in tractor drivers, as a whole collection of individual factors usually contribute. Whole-body vibration stress alone would be insufficient to explain the increased morbidity of stomach complaints in this working group.

Thus, the epidemiological study of the team from Gorki (1975), who discovered that the highest proportion of stomach illnesses was to be found in taxi drivers (with a comparatively low vibration exposure), showed that the source of higher rates of stomach illnesses in drivers subjected to vibration stress cannot necessarily be ascribed exclusively to vibration stress. In drivers of vehicles and heavy construction machinery and tractors under multiple stress (noise, unfavorable climatic conditions, work shifts, unphysiological diet conditions, strong psychological stress, etc.), it can instead be concluded that, etiologically speaking, there are multifactorial contributory causes.

The factor of "whole-body vibration" must nevertheless play a large role in the etiology of stomach diseases. Because of vehicle vibration, with frequencies that often lie in the region of the resonance frequency of the stomach, the mechanical strain on the stomach must be regarded as high. In the opinion of Rosegger and Rosegger (1960), long-term exposure can result in a change in the reaction of the vegetative nervous system with a secondary effect on the activity of the stomach-intestinal canal. They believed that the increased morbidity is a result of the shaking of the stomach and of the change in tone of the smooth musculature of the stomach arising from the change in reaction of the vegetative nervous system. Many authors, therefore, ascribe strong vibration stress as *one* source of stomach disorders.

Undoubtedly, a predisposition toward stomach illnesses can considerably favor its occurrence, particularly if there are other exogenous factors such as drug and alcohol abuse, as well as too much rich food, psychological stress, etc. The therapy of such "vibration-induced stomach diseases" must also take these factors into consideration, in addition to the vibration stress.

7. The Influence of Particular Modes of Vibration

7.1 Vibration Occurring Simultaneously in Several Directions

When vibration stress occurs at the place of work, 1, 2, or 3 directions of the coordinate system related to man, or vectors of these axes, can be involved (Fig. 1). In such multiaxial vibration, the stress may predominate in one direction or may have similar magnitudes in all directions. In addition, rotational vibration about these axes may also take place (see section "Rotational Vibration," p. 125).

In laboratories, the acute effects of mechanical whole-body vibration under the influence of stress are almost exclusively investigated using only one vibration direction. Only occasionally has research been conducted on the effects of the stress in two axes (Dupuis et al. 1974; Griffin and Whitham 1977).

However, it can be maintained that all field investigations of the chronic effects of mechanical whole-body vibration involve, as a rule, vibration stress in several axes, as occurs, for example, in earth-moving machines and agricultural and forestry tractors. To date, however, there have been no systematic studies of the effect of vibration in different axes or more than one axis in order to assess the impairment of well-being, performance, and health.

In regard to this question, Griffin and Whitham (1977) mentioned that experiments have been carried out by Holloway and Brunigham (1972) and Kirby et al. (1975). However, Griffin and Whitham believed that the experiments were insufficient and therefore conducted their own, using vibration stress on two axes in the directions, y and z, at 3.15 Hz. They showed that the sum vibration perception is greater when the stress is in several directions than when the stress is caused by vibration in only one direction. On the basis of their experimental results, they therefore recommended that if the evaluation of vibration stress, based exclusively on the maximum directional components, is sufficient, then the "vector sum" should be determined, i.e., the square root of the sum of squares of the frequency-weighted vibration stress in the different directions. Their suggestions correspond with the findings of Hansson and Wikström (1981) from field experiments. However, they also admit that further research is required in order to resolve this question.

Because of the lack of any other data, the primary recommendation here is that vibration stress be determined as the K value or the weighted acceleration value a_w in each direction (corresponding to VDI 2057 or ISO 2631). To predict the risk to well-being, performance and health, the greatest stress of the three vibration axes should be compared with the recommended levels of these standards, as they refer to the measurement and evaluation of vibration in *one* direction only. If there is also high stress in other vibration axes, then the situation should be taken particularly seriously – without, however, venturing to provide further concrete data on the potentially increased risk.

As mentioned previously, "vectorial summation" of stress from several directions can be useful to compare observations of vibration stress in different machines, places of work, or vehicles. However, there is insufficient knowledge at present to enable occupational medicine to use such values to evaluate the risk to health because, as mentioned previously, VDI 2057 and ISO 2631 only refer to vibration stress in *one* direction.

7.2 Mixtures of Periodic Vibrations, Random Vibration, and Vibration Containing Shocks

Since by far most of the experiments reported in the section "Acute Effects of Mechanical Vibration," p. 12, were under sinusoidal vibration, the question arises as to whether or not the findings are also valid if the stress comes from complex periodical (meaning regular) or from random (irregular) or even vibration containing shocks. In contrast, the chronic occupational effects of mechanical vibration arise, as a rule, from the influence of random vibration, e.g., in vehicles.

Of those experiments carried out since the late 1960s, the following should be mentioned as they dealt with the effect of mixtures of two or more sinusoidal types of vibration: Miwa (1968a), Lange (1971, 1974b), Dupuis et al. (1972), Fothergill (1973), and Fothergill and Griffin (1977a). The following experiments took the effect of random vibration into consideration: Miwa (1969), Dupuis et al. (1972), Dupuis et al. (1974), Griffin (1976b), Bastek et al. (1977), Donati et al. (1982), Moseley et al. (1982), Hansson and Wikström (1981). On the other hand, only a few studies have been concerned with the effects of repeated shocks (Miwa 1968c; Allen 1976; Griffin and Whitham 1980).

7.2.1 Mixtures of Periodic Vibration

Lange (1971, 1974b) investigated the subjective evaluation of mixtures of sinusoidal vibration. He found that if the mixture contained frequencies lying in the main resonance region for man (about 5 Hz), then the evaluation was dependent entirely on them. At fundamental frequencies below 5 Hz, the second and third harmonics were of considerable influence, while at fundamental frequencies of 5–8 Hz, harmonics higher than the second had hardly any influence on perception.

With regard to perception of one component in a mixture of two types of sinusoidal vibration, Dupuis et al. (1972) found that there is a "masking effect," such as in acoustics, for vibration of > 2 Hz. Thus, the component over 2 Hz is detected at a greater intensity than would be the case for pure sinusoidal vibration. The opposite occurs in mixtures for vibration of < 2 Hz – the component below 2 Hz is detected at a lower intensity.

The studies by Fothergill (1973) and Fothergill and Griffin (1977a) have contributed a great deal to simplifying the evaluation of vibration inasmuch as they have established that periodic vibration mixtures such as these can be best evaluated by means of modifying the acceleration spectrum using an "electronic frequency-weighting network." Such a weighting network must correspond to the frequency-dependent effects on man and was recommended in VDI guideline 2057 in 1963. To obtain practically comparable results when carrying out third or octave analyses rather than using a weighting network, the value of each frequency band is squared, summed, and finally the square root is obtained. With this method Hansson and Wikström (1981) found a better correlation to subjective vibration perception than with the "rating procedure" originally recommended in ISO 2631.

In contrast to this rating procedure, which takes into consideration only the frequency band that exceeds the highest evaluation curve, the "weighting procedure" has always been recommended in VDI 2057. The weighting method has been proved successfully by the above-mentioned studies and others, and thus became the foundation for the German standard (VDI 2057) and the international standard (ISO 631), Amendment 1, 1982, for the evaluation of vibration.

7.2.2 Random Vibration

Dupuis et al. (1972) measured the vertical vibration in small trucks, wheeled and tracked tractors in field experiments and reproduced them on a vibration simulator. Subjects were exposed to this random vibration to compare it with sinusoidal vibration in the vertical z direction. In all cases, the subjects showed greater sensitivity under the influence of random vibration, as opposed to sinusoidal vibration. This was especially apparent in the resonance region of the trunk, but this high sensitivity decreased with increasing frequency. These data correlate with the recent research results of Donati et al. (1982), who extended the results to vibration stress in the x and y directions.

Dupuis et al. (1974) have investigated 11 different physiological and psychological criteria of strain from the effect of random truck vibration in the z and x directions. Vibration-induced changes showed themselves above all in flicker-fusion frequency, tracking performance, and visual perception time (tachistoscopy).

A research team from the German Democratic Republic and the USSR (Bastek et al. 1977) compared the effects of sinusoidal vibration and random vibration on biodynamic behavior, the electrical activity on the dorsal muscle groups, equilibrium regulation, and on psychophysical performance parameters. They found approximately similar biological reactions to the serious types of vibration stress, which however also pin-pointed the marked effects of random vibration stress. Dupuis et al. (1972) were able to demonstrate these marked reactions very clearly in the muscle activity of the m. erector spinae during random vibration.

From his research results, Griffin (1976 b) concluded that the weighted acceleration of a random broadband spectrum vibration represents a suitable criterion for the evaluation of the subjective effect of irregular vibration, as long as the vibration does not contain large shock components.

Finally, Moseley et al. (1982) found, when comparing sinusoidal and random vibration, that changes in visual perception – in terms of reading errors – can be fairly well predicted and therefore evaluated by using the root-mean-square value of the frequency-weighted acceleration for both types of vibration.

7.2.3 Vibration Containing Shocks

The shock content of vibration can be extremely varied. A single shock of short duration, without repetition, but with a high acceleration peak, can hardly be compared with continuous vibration (Kleinhanss and Dupuis 1971; Dupuis and Kleinhanss 1973; Dupuis 1981 c). At the very least, repeated shocks in mixtures of vibration appear to be subjectively more strongly perceived (Yonekawa 1975), although few data on this problem are available. The shock elements of random vibration vary in terms of frequency of occurrence and magnitude of acceleration (which is mostly expressed as the "crest factor," the ratio of peak value to root-mean-square value of acceleration). However, the crest factor yields practically no data on the frequency at which such peak values occur. Furthermore, it is an open question whether the effect on man is the same when a high crest factor occurs in conjunction with a very low or very high root- mean-square value. Apart from the above-mentioned isolated experiments, such as Miwa (1968 c) and Griffin and Whitham (1980 a), to date there has been no systematic research on vibration containing shocks, and there are also no well-founded findings on the evaluation of vibration that differ from ISO 2631 and VDI 2057. According to the present state of knowledge, one can presume that from the point of view of chronic injuries, vibrations with weighted acceleration crest factors of ≤ 6 probably lead to no stronger biological reactions, whereas those of > 6 may have stronger effects compared with nonshock-type vibration.

To summarize the research results already presented, it can be assumed that mixtures of periodic vibration, random, and moderate repeated shock vibrations with crest factors of ≤ 6 can best be evaluated from the viewpoint of the effects to be expected, with the K value or the root-mean-square value of the frequency-weighted acceleration a_w. However, it remains still unanswered as to whether strong repeated shock vibration has comparatively stronger effects and thus must be more strictly evaluated with some correction factor. Additional studies are therefore still necessary as regards the question of the effect of vibration containing shock.

7.3 Rotational Vibration

The effects of mechanical vibration described so far have dealt almost exclusively with translational, that is, unidirectional vibration, in one of the three coordinate axes, x, y, and z, relating to man (Fig. 1). Only occasionally have the effects of rotational vibration about one of these three axes been mentioned, as for example in the discussion on kinetosis (section "Kinetosis," see p. 66). There is no doubt, however, that at various places of work and in certain means of transportation (e.g., ships, airplanes, helicopters, self-driven earth-moving equipment with a short wheel base), rotational vibration in addition to translational vibration plays a role in the stress on man. Therefore, the effects on man established in epidemiological studies must be attributed to the total translational and rotational vibration stress, although rotational vibration, as a rule, is not measured in field studies.

Only in the last 10 years has human strain under stress from rotational vibration been researched, and there have been relatively few studies. These studies have dealt with the influence on performance (Sjøflot and Suggs 1973; Rühmann 1978), and above all psychophysical reactions, i.e., subjective perception in the sitting position (Pradko 1965; Simic 1970; Sjøflot and Suggs 1973; Parsons and Griffin 1978a, 1978b; Parsons et al. 1979; Ilgmann 1979; Shoenberger 1979, 1980) and in the recumbent position (Helling 1978; Irwin 1981).

Sjøflot and Suggs (1973) investigated the effect of roll vibration about the horizontal x axis, where the axis lay in the area of the seat base. Their assessment measures − heart rate, control errors, and subjective evaluation − showed joint effects that, in general, increased from pure vertical vibration through roll motion to combinations of roll and vertical vibration.

According to the work of Pradko (1965) and Parsons and Griffin (1978a), the sensitivity to roll diminishes as the frequency increases from 1 to 30 Hz. Also, Shoenberger (1979) found the same tendency between 2.5 and 10 Hz and Ilgman (1979) between 1 and 10 Hz, and at various acceleration intensities.

Thus it is clear that there are no frequency-dependent maxima of sensitivity arising from body resonances for such rotational motion, as is the general case for translational vibration.

As regards the effect of pitch vibration, the results of Parsons and Griffin (1978a) agree with those of Simic (1970) and support the same frequency dependence, i.e., reduced sensitivity with increasing frequency. For the same rotational acceleration, there was lower sensitivity to pitch than to roll.

Yaw vibration about the vertical axis would seem, from their appearance, to have minimal practical significance. Also, the effect on man can be expected, according to the investigations of Shoenberger (1980), to be significantly less than that for pitch and roll. His subjects, who were using a seat harness, chose the least rotational acceleration in pitch, higher in roll, and the highest rotational acceleration in yaw for equal intensity at frequencies 2.5 and 8.0 Hz.

In the reclining body position, such as occurs in the transport of the sick and injured, Helling 1978) has investigated rotational vibration about the horizontal y axis and established frequency-dependent curves of equal perception which, in this case, indicated a particularly high sensitivity at around 30 Hz.

Furthermore, from the studies of Parsons and Griffin (1978b), it can be deducted that for rotational vibration, with increasing distance from the rotational axis, the subjective intensity is more related to translational motion as the largest vector component of the vibration signal.

The question now arises as to whether stress from rotational as well as translational vibration must be taken into consideration when evaluating a situation and, if so, how. The relatively few studies presented so far do not yield generally valid data and thus to date there are no corresponding recommendations in VDI guideline 2057 and ISO standard 2631.

As pointed out by Parsons and Griffin (1978), for certain kinds of stress from translational and rotational vibration components, it is sufficient to use the highest frequency-weighted translational component to evaluate adequately the extent to which feelings of well-being are detrimentally affected. In this connection, the field experiments of Parsons et al. (1979) have shown that in vehicles, translational vibration, especially in the vertical direction, generally leads to more discomfort than does rotational vibration.

Because of the lack of further results, it must be recommended that rotational vibration should be subdivided into its translational components, measured and evaluated in accordance with VDI 2057 or ISO 2631.

7.4 Exposure Duration and Rest Pauses

Two questions arise with regard to the effect of the duration of exosure to vibration. First, does the increase of exposure from minutes to hours to a complete working day result in greater and greater effects of vibration on psychophysiological well-being and performance?

There are only a few experimental data on this question. Thus, in the course of his 2-h experiments on simulated tractor-vibration stress (with $K = 27.5$), Christ (1973) determined that heart rate (after a short initial climb) and flicker-fusion frequency decreased, as a result of monotony and a time-dependent adaptation process. These data are valid both with and without simultaneous tracking tasks. During these 2 h, the average steering errors in this task remained unchanged. In addition, Gray et al. (1976), using 2-h stress (sinusoidal vibration 5 Hz, $a_{eff} = 1.2$ m/s^2, corresponding to $K = 24$), found "little proof" of a decrease in performance as a result of duration-dependent exposure. This team used an audio vigilance test, an alphabet letter search test, a tracking test, and a writing test. Moreover, Scheibe (1979) showed that individual reaction behavior can vary; he used two subjects in stress experiments of up to 120-min duration. Under vibration stress, one subject showed a time-varied decrease in heart rate and a decreased or constant muscle strain with simultaneous, monotonous reduction of steering performance. In contrast, the other subject showed an increase in central and peripheral signs of strain while maintaining time-dependent ongoing performance. In the latter subject, there was apparently a higher rate of burnout regarding performance as a result of increasing psychophysiological stress with exposure time.

The experiments by Scheibe (1979) resulted in data from a greater number of subjects with regard to the effects of frequency and acceleration of vibration on subjectively tolerated duration of exposure of 10−240 min. These data correlate with the "fatigue-decreased proficiency boundary" of ISO 2631, i.e., decreasing frequency-weighted acceleration with the increase of daily duration of exposure. The experiments of Seidel et al. (1980), using 3-h vibration exposure, showed that above this boundary there were undesired effects and decreases in performance.

The second question that arises, from the point of view of duration of exposure, pertains to whether and to what extent increasing vibration intensity demands reduced daily exposure time to keep to the same risk to health. However, there are apparently no systematic experiments on this. The risk of vibration-induced effects on health is undoubtedly a dose−effect problem in which the concept of a "dose" includes the K value or frequency-weighted acceleration a_w and the duration of exposure.

In guidelines for occupational medicine, duration of exposure is, first of all, expressed only in terms of the daily period of vibration stress, whereas there is no information concerning the duration of exposure during professional life that indicates a certain risk. However, from experience it can be presumed that only after several working years is there risk of vibration-induced injury (see section "Chronic Effects of Whole-Body Vibration," p. 87).

If the vibration intensity is high, this must be balanced by a shorter duration of daily exposure; if the intensity is low, a higher daily exposure will result in the same risk. Relevant evaluation curves are given in ISO standard 2631 for the first possibility of health risk. These curves have been accepted for VDI 2057. They correspond to about half of the vibration intensity that represents the pain threshold for healthy males. A peak acceleration value of 9.8 m/s², corresponding to a root-mean-square value of 7.1 m/s², should never be permitted since then, if the person is not wearing a seatbelt, he will be lifted from the seat or platform.

In accordance with previous practical experience, there is little reason to assume that the evaluation values represented in VDI 2057 and ISO 2631, regarding the risk of impairment dependent on the duration of exposure, are incorrect (see section "Laws, Regulations, Standards and Guidelines for the Protection of Man against Mechanical Vibration," p. 133). Today there are tendencies to straighten the time-dependent evaluation curves for easier use (Griffin and Whitham 1980b; ISO 2631, Amendment 1, 1982).

Christ (1973), Dudek et al. (1973), and Scheibe (1979) have conducted experiments on the influence of pauses, i.e., interrupted vibration exposure. These experiments have shown that the physiological parameters, heart rate and flicker-fusion frequency, are dependent on time-dependent adaptation processes, which are interrupted during each pause (Christ 1973). The time-dependent fatigue processes of muscle activity were also interrupted by the breaks in exposure, however. Thus, pauses in vibration − analogous to the effect of short breaks during hard work − are considered to have a recuperative effect, which is even more enhanced when there are many short pauses rather than fewer, long pauses (Scheibe 1979). It can be expected that the recuperative effect from physical fatigue during high vibration stress is particularly noticeable.

In part 3 of VDI recommendation 2057 (1979), Note 7 says with respect to this question: "Practical experience has shown that when vibration intensity is high, performance is influenced less if the time of vibration exposure is interrupted by pauses distributed throughout the work shift. Also, as is shown in practical experience with occupation-related vibration stress, the danger to health is less if the duration of the exposure is interrupted by short breaks. However, there is insufficient knowledge so far to acquire quantitative data on the effect of pauses."

8. Preventive Measures

As a rule, it is only possible to protect people effectively from mechanical vibration by means of a combination of various measures (Schäfer et al. 1982). As with protection from noise, all preventive measures can be divided into four main groups:
− Engineering methods
− Prevention by work organization
− Personal prevention
− Prevention by occupational medicine

8.1 Engineering Methods

Most important is an appropriate selection of equipment, machines, tools, and procedures for use and the careful maintenance and upkeep of the equipment.

Further, one should differentiate between constructive measures to reduce vibration at source and reduction in transmission of vibration. To avoid or reduce vibration at source, a change in working principles is sometimes appropriate, such as rotating instead of reciprocating machine parts, belts instead of chains, drilling instead of stamping, hammering, or ramming, pressing instead of beating, gluing instead of riveting. If such procedures, however, cannot be considered because of other technical, power, or economic disadvantages, further possibilities must be sought to reduce vibration at its source. Thus, the number of revolutions and the speed of moving machine parts must be chosen in such a way that the resulting vibration frequency is not in a range to which man is particularly sensitive. The removal of unbalanced rotating masses with subsequent mass compensation, the coupling of additional masses to shift the natural frequency of the machine, and the use of inertial masses can all be suitable measures (DIN 3831). The welding of tracks for tracked cranes also contributes to the reduction of vibration stimulation, as does the regularly used graders to level roads used by earth-moving equipment (Schäfer et al. 1982).

Very often measures are taken to reduce vibration transmission. As a rule, such measures are very economical and can be realized, for example, by means of vibration isolation elements (springs, dampers, isolating plates) or by suspension seat systems. In these seat systems the spring and damping devices must be very carefully balanced to match the vibration sources in the vehicle or ma-

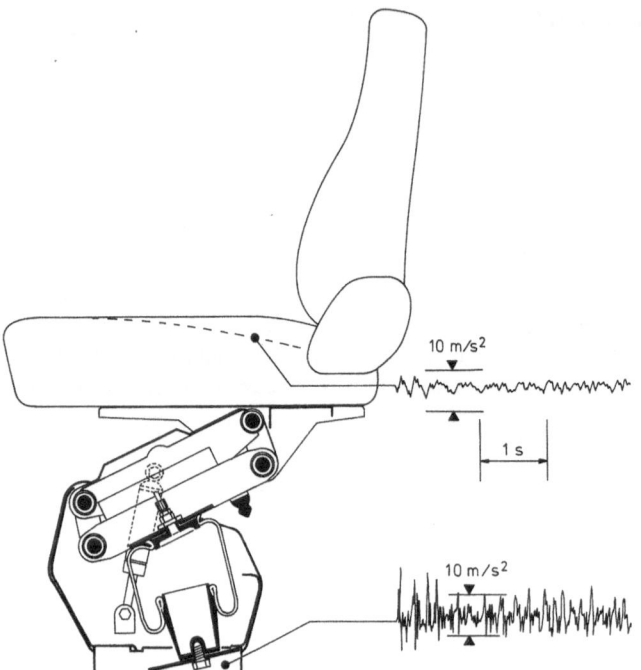

10 m/s²

1 s

10 m/s²

Fig. 51. Reduction in vibration by means of pneumatic seat suspension

chine. Adjustment devices should take the weight of the driver into consideration. These measures can result in a considerable reduction in vibration (Dupuis et al. 1982b; Köhne et al. 1982), as shown in Fig. 51 (example of an air-suspended driver's seat).

8.2 Prevention by Work Organization

The work-organization preventive measures taken against vibration-based degradation in well-being, impairment of human performance, and danger to health are all basically aimed at a reduction in the duration of the effect of mechanical vibration.

The procedure employed so far to reduce the daily duration of exposure to whole-body vibration on man, as recommended in part 3 of VDI guideline 2057 (1979), is especially justified when the vibration stress is high. The procedure does not necessarily result in a general reduction of vibration exposure, unless the number of years in the occupation and, hence, the total number of hours' exposure of a working life are taken into consideration at the same time. If need be, limitation of exposure during a working life must therefore also be considered.

From the point of view of occupational medicine, pauses in vibration exposure should also be recommended (see section "Exposure Duration and Rest Pauses, p. 126). Pauses, as long as they are connected with an interruption of the remaining work stressors (noise, unfavorable climatic conditions, etc.), counteract the impairment of well-being and/or performance by means of the recovery effect. Organizationally, these vibration pauses are accomplished in such a way that a pause of at least half an hour should be established, for example, after 1-h occupation activity involving whole-body vibration stress. During these pauses the person normally under the effect of vibration can carry out other activities, e.g., maintenance work.

Finally, after a certain period of exposure, a change in the work place should be discussed from the viewpoint of job rotation.

8.3 Personal Prevention

The possibility of personal prevention against vibration-induced diseases or injuries (i.e., the use of passive protective measures in the sense of body protection to reduce vibration) are extremely limited in tractor drivers, drivers of heavy construction equipment, truck drivers, etc.

Low-frequency air vibration (infrasound) is disseminated in the human body in the form of mechanical vibration and, depending on the frequency, can stimulate certain organs to resonance: Sound-protection suits are therefore useful, except for low-frequency mechanical vibration, which has considerable displacement amplitudes and therefore cannot be protected in this way. According to roentgenocinematographical investigations, the use of abdominal belts has also not proven useful in limiting the extent of vibratory movements of the stomach (see section "Vibration Behavior of Internal Organs," p. 39).

Compensatory exercise regularly carried out by exposed workers on an individual and voluntary basis can be included as personal preventive measures and may take place in rooms set aside by the employer. A specific compensatory sport can also serve the same purpose, generally strengthening the trunk musculature.

8.4 Prevention by Occupational Medicine

Special emphasis should be placed upon prevention using an occupational medicine approach to be able to estimate the risk of disease as early as possible, especially from the point of view of the performance ability of the spinal column. Thus, various authors have expressed a belief that initial and follow-up examinations are important from the point of view of occupational medicine (Kunz and Meyer 1969; Franke 1978; Kühne et al. 1982). (See section

"Methodological Problems," p. 87, and "Medical Examinations of New Employees and Regular Health Check-ups," p. 138) Junghanns (1979) feels it is a great pity that suggestions to pay more attention to the spinal column when examining new employees have not yet found the necessary response. At present, "Occupational Health Regulations for Prevention Against Whole-Body Vibration" are in preparation in Germany for physical examinations.

In the initial examination, in addition to a general examination, a thorough work history should be taken and special examinations carried out, e.g., of the spinal column and stomach. If necessary, radiographs of segments of the spinal column must also be included.

The following medical findings are to be considered as disqualifications for people to be employed under exposure to whole-body vibration:
− Maldevelopments of the spinal column (spina bifida, etc.)
− Arrested Scheuermann's disturbance in growth with incomplete healing
− Pronounced distortion and/or stiffening of the spinal column
− Recidivistic, painful impairment of function of the spinal column
− Clear degenerate processes of the spinal column, e.g., spondylosis deformans, spondylostheochondrosis, uncovertebral spondylosis, spondylolisthesis, spondylolysis
− Systemic inflammatory diseases of the spinal column or joints
− Vertebral processes that have been operated upon
− Old fractures of the spinal column (also fracture of the spinal process and transverse process)
− Chronic stomach and duodenal diseases, e.g., gastritis
− Partial resection of the stomach, vagotomy
− Pronounced degenerative joint diseases or inflammatory joint diseases and deformation of the lower extremities (only in the standing position)

It seems advisable to carry out follow-up examinations in workers up to 50 years of age every 4 years, and in those over 50 years of age every 3 years. The exclusion criteria described for initial examinations should also take into consideration whether there are any permanent health risks that rule out continuation of the activity involving whole-body vibration.

Physiotherapeutic measures have proved to be worthwhile in the prevention of complaints or injuries of the spinal column in workers exposed to vibration. The company physician or general practitioner should not be too narrowly restrictive in regard to indications for this kind of physical therapy with its variety of methods. As a further preventive measure, regular treatment should be considered, with stress on spinal or stomach problems.

Finally, the possibility of early retirement should also be mentioned whenever the worker has been under the effect of whole-body vibration for a very long time.

9. Laws, Regulations, Standards, and Guidelines for the Protection of Man Against Mechanical Vibration

Regulations that serve to protect workers from the dangers arising from mechanical vibration have been established by the legislators and agencies responsible for work protection, such as ministries of labor and social affairs and industrial insurance organizations. Furthermore, international agreements have also established such protective measures. The corresponding laws, rules or agreements relate to:

1. Occupational diseases
2. Criteria for the evaluation of risk from mechanical vibration and the establishment of exposure limit values
3. Guidelines for the measurement and evaluation of vibration
4. Medical examinations before employment and regular health examinations
5. Technical and organization measures for protection against mechanical vibration

9.1 Occupational Diseases

Included in the list of occupational diseases in international agreement No. 121 of the ILO (1980) is No. 23: "Diseases caused by vibration disorders of muscles, tendons, bones, joints, peripheral blood vessels or peripheral nerves."

The international experts who drew up the list of occupational illnesses also recognized "the occupational origin of diseases of the spinal column resulting from vibration of the whole body and from shaking, especially in drivers of power-driven machines and tractors," although they admitted that complaints in the area of the lumbar spine are not typical (specific) and that their origin is therefore difficult to prove.

On a national basis (in Germany), occupational diseases from "whole-body vibration" are not mentioned in the latest version of the list of occupational diseases (BK list), which modifies the 7th regulation on occupational diseases of 8 December 1976. However, the legislators have created a general clause for single cases in § 551, section 2, of the national insurance regulation (RVO). This regulation also provides compensation for diseases that are not contained in the BK list as "occupational diseases." This regulation also provides compensation for exceptions, with the goal of helping in cases where the list of occupational diseases is not always up to date as regards the latest technical developments and scientific findings.

Fulfillment of the following requirements is necessary for use of § 551, section 2:

- A certain group must be exposed in their occupational work to particular stress to a much higher degree than the remaining population.
- This stress must be designated as causing diseases of this kind, according to the latest results of medical science.
- These medical results must not have been sufficiently well established or not yet proved in the previous version of the occupation diseases BK list.
- The causal relationship between the occupational stress and the diseases must be concretely probable.

It is obvious here that a review, in terms of sufficient probability of a causal relationship, will demand in each individual case not only an exact clinical and radiological clarification, but also a precise work history of professional exposure to vibration.

9.2 Criteria for the Evaluation of Risk from Mechanical Vibration and the Establishment of Exposure Guidelines

According to ILO Agreement No. 148, Part III, Para. 1 (1977), there is international agreement that the national "competent authority shall establish criteria for determining the hazards of exposure to vibration in the working environment and, where appropriate, shall specify exposure limits on the basis of these criteria".

In the Federal Republic of Germany there are still no papers or preventive regulations regarding whole-body vibration. However, the preventive regulation "Vibration" is in preparation, in which instructions will be given as regards the occurrence and sources of risk, as well as measures for protection from vibration. With one exception for agricultural tractors, there has still been no determination of exposure limit values, but only general guidelines. Thus Section (10) of the guidelines for the construction and equipment of driver cabs in power-driven vehicles, tractors, and work equipment (driver cab guidelines) of 16 December 1966 requires that:

The seat must be sufficiently suspended, upholstered, and damped. A special damping component is not required if the purpose is thought to be attained in other ways.

Section (31) establishes:

Noise and mechanical vibration: the noise and mechanical vibration affecting the driver in the driving cab or driver's space is not allowed to surpass what can be reasonably avoided, according to the present status of technology.

VDI Guideline 2782 "Recommendations for the construction of vehicle driver seats in power-driven vehicles" of April 1971, as well as VDI 2783 "Recommendations for the construction of front and back passenger seats" of

November 1972, both include the aforementioned specification from the driver cab guidelines in the paragraph "Vibration isolation." However, this text has been expanded as follows:

No generally binding limit value is available with regard to physiologically acceptable vertical acceleration at the seat surface (see, however, VDI 2057, which gives provisional evaluation values).

The German Agricultural Injuries Insurance Institute has a work-protection regulation for agricultural vehicles that has been valid since 1 January 1970. This regulation includes a testing procedure developed by Dupuis and Hartung (1966) for the vibration behavior of tractor seats, the highest threshold of which permits a K value of no more than $K = 25$. (This corresponds to frequency-weighted rms acceleration $a_w = 1.25$ m/s^2.) In the meantime, a similar testing procedure has been standardized for the countries in the European Common Market and a frequency-weighted acceleration $a_w = 1.25/s^2$ (rms, which corresponds to $K = 25$) has been determined as the limit value (EWG 78/764, 1978; EWG 83/190, 1983).

For earth-moving equipment, the work-protection regulations of the Injuries Insurance Institute for Underground Construction Engineering (1976) in § 7, Chap. 1, so far require only a general application: "The driver's seat in earth-moving equipment must be adjustable and constructed in such a way that it has a suspension system and shock absorption so that injuries to health from vibration will be avoided." It must be taken into consideration, however, that in the foreseeable future, the standard DIN-ISO 7096 (1982) will also be applied for certain earth-moving machines, as the limit value in this standard is the same as the aforementioned guidelines for agricultural tractors of the European Common Market.

With regard to certain groups of people, such as pregnant women and young people who have a particular need to be protected from mechanical vibration, there are particular regulations.

9.3 Guidelines for the Measurement and Evaluation of Vibration

For any evaluation of protective measures against mechanical vibration and any assessment of work history, the vibration at the place of work must be measured and evaluated. There are available surveys on methods of measurement and analysis concerning mechanical vibration at the place of work (Dupuis 1981 a), as well as the current status of the medical evaluation of mechanical vibration at the place of work (Dupuis 1980 a).

Accordingly, international standard ISO 2631 (1978) and national guideline VDI 2057 can be applied for the measurement and evaluation of mechanical vibration, particularly in construction industry German standard DIN 4150 (1975). These guidelines correspond to current knowledge in occupational medicine.

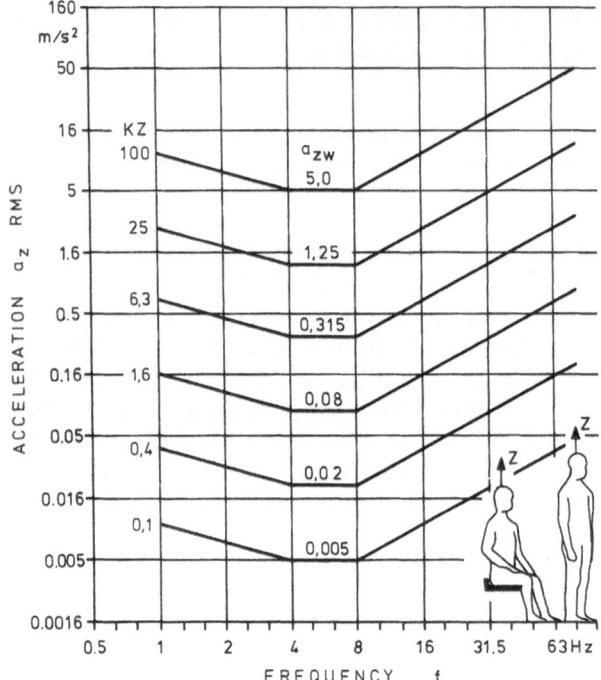

Fig. 52. Frequency-weighting curves for vibration in z direction (seated and erect body posture)

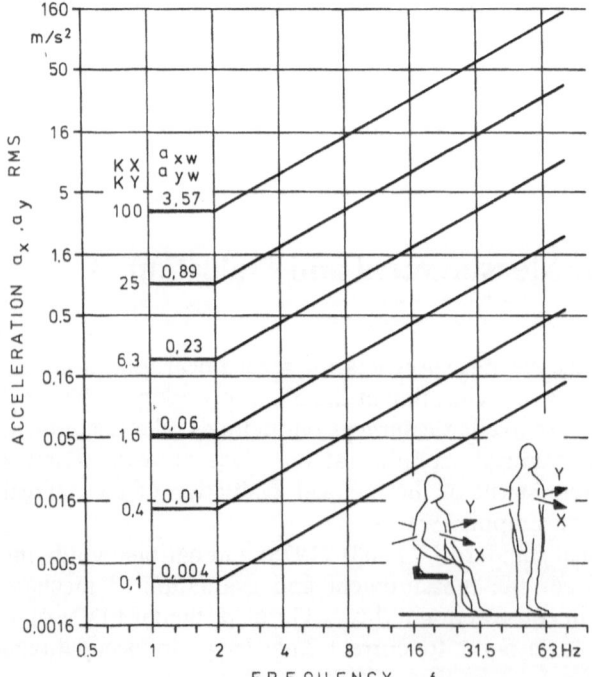

Fig. 53. Frequency-weighting curves for vibration in the x, y direction (seated and erect body posture)

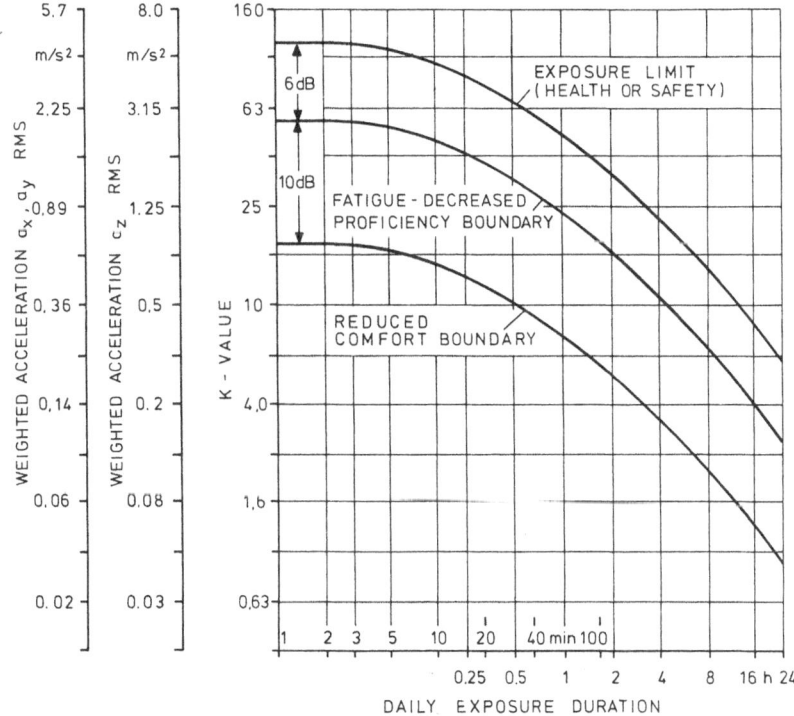

Fig. 54. Assessment of daily vibration exposure, depending on weighted acceleration rms

For the seated and standing body posture, the frequency weighting is the same, but there are differences, however, for the vertical vibration direction, z, on the one hand (Fig. 52), and for the horizontal vibration directions, x and y, on the other (Fig. 53). Nevertheless, other evaluation curves are valid for the unspecified body postures, such as the freely moving person in an environment where there is vibration (dwelling in buildings, on board ships). For the reclining position (e.g., transport of the sick), other evaluation curves are provided in VDI 2057.

The evaluation of the derived K value or a_w value is based on the criteria of comfort, performance, and health. The curves contained in this guideline (Fig. 54) are not to be understood as "limit values" but as "guide values," and if these values are surpassed, a degradation in comfort, performance, or health has to be taken into account. However, there are still no quantitative data on the magnitude of risk. At present, the following influences cannot be taken into consideration because of the lack of sufficient knowledge: the effect of single shocks and vibration-containing significant shocks; the simultaneous effects of vibration in more than one direction; the effect of pauses in exposure.

9.4 Medical Examinations of New Employees and Regular Health Checkups

In Article 11 in International Agreement No. 148 of the ILO (1977) it is established:

There shall be supervision at suitable intervals, on conditions and in circumstances determined by the competent authority, of the health of workers exposed or liable to be exposed to occupational hazards due to vibration in the working environment. Such supervision shall include a pre-assignment medical examination and periodical examinations, as determined by the competent authority.

However, the ILO Agreement has not yet been included in German law in the Federal Republic of Germany, as at present there are still no corresponding regulations for such health examinations.

The German Central Association of the Industrial Injuries Insurance Institute has, however, assigned its committee "Occupational Medicine" the task of Working group 2.2 "Vibration" for clarification. Experts are to be used to standardize methods for use in diagnostic examinations and to determine which groups of persons are to be examined. This committee report should finally lead to occupational health regulations for prevention against "whole-body vibration" (see section "Prevention by Occupational Medicine," p. 131).

9.5 Technical and Organizational Measures for Protection Against Mechanical Vibration

According to ILO Agreement No. 148, Article 9 (1977):

... the working environment shall be kept free from any hazard due to vibration –
(a) by technical measures applied to new plant or processes in design or installation or added to existing plant or processes; or, where this is not possible,
(b) by supplementary organisational measures.
Where the measures taken in pursuance of Article 9 do not bring ... vibration in the working environment within the limits specified in pursuance of Article 8, the employer shall provide and maintain suitable personal protective equipment. The employer shall not require a worker to work without the personal protective equipment provided in pursuance of this Article.

To date, there are still no regulations concerning organizational protective measures which, for example, could take the form of the establishment of maximal daily exposure times and the planning of rest breaks. Except for the limitations established for daily travel time for test drivers in the armed forces, there are still no regulations on the use of personal protection devices.

10. Summary and Conclusions

Protective measures against mechanical whole-body vibration require knowledge of not only the kind and extent of stress that occurs but also the effects on man. Current knowledge of the vibration stress that occurs in the place of work has been improved in the last 10 years by research investigations carried out under practical field conditions.

Particularly since the 1950s, experiments have been conducted in many countries on the effects of such vibration on man as regards acute and chronic strain. Thus, it has also been possible to improve current knowledge from the point of view of occupational medicine.

First of all, the biodynamic vibration behavior of the exposed body, including its parts and organs, is important. In this respect, there are comprehensive data from which the frequency-dependent transmissibility functions with the essential resonance regions can be deducted, including the influence of various body positions and stimuli in different directions. However, the existence of resonance always means particularly high strain for the body tissue involved.

The changes in physiological functions pertain to the circulatory system in only a very limited way. Thus, an increase in heart rate is clear only when the vibration stress is very high, blood-pressure changes are not uniform, and ECG changes cannot be significantly differentiated. There has been very little research on peripheral disturbances in blood flow in the sense of vaso-constrictions, and these disturbances are apparently considerably less and have less effect than in hand-arm vibration, in which the disturbances are markedly influenced by the gripping forces in the fingers.

From the point of view of respiration, hyperventilation as a result of vibration has been proved. It is particularly noticeable in the resonance range for the human body and at high accelerations. As respiration is more strongly influenced than to be expected from the increased energy requirements of vibration, hyperventilation is only energetically conditioned in a small way. Instead, it may be caused by passive movements of the diaphragm and the abdominal wall, arising from motion of the abdominal contents and leading to a kind of artificial respiration, whereby ventilation of the dead space is apparently primarily involved.

Muscle activity is always significantly increased under vibration stress. This can be explained by the defensive reaction of the main muscle groups which undertake a protective function.

Concerning the vegetative reactions, there are indications that when under stress from whole-body vibration, the natural muscle reflexes controlled by the peripheral nervous system can be reduced. Vibration-induced changes in the electroencephalogram, the blood count and the endocrine system, as well as biochemical changes have to date not been proven to be generally valid. Corresponding changes instead appear to lie in the region of the physiological norm.

The question regarding whether mechanical vibration can cause deafness or lead to hearing difficulties has so far not been clearly answered. The fact that some of the experimental results have varied to such a degree can only be explained by the vibration stress intensities chosen by the authors, which have differed greatly as regards sound intensity and sound spectrum, acceleration amplitude and frequency, and exposure time. Thus, no generally valid and transferable data are available so far.

However, the vestibular system is clearly affected — above all during rotational vibration. In low-frequency vibration, kinetosis can be brought about, and there is still little knowledge on this, particularly in relation to the kind of vibration stress, the other stress factors involved, and the adaptation procedures.

The influence on visual perception has been fairly thoroughly investigated, but because of the many kinds of influence factors, these influences are not yet completely understood from the viewpoint of genesis. Visual perception time is particularly impaired. Visual acuity is reduced by resonance vibration of the eyeball.

With regard to performance during exposure to low levels of vibration stress, the fatigue effect can be compensated by corresponding motivation. As the intensity of vibration climbs, however, a decrease in performance must be taken into account. Because of the variety of tasks, the construction of operating controls and individual requirements, it is not possible to provide absolute limit values under which no impairment can be expected. An especially high impairment of performance can be expected during vibration stress which results in high transmission factors to the human body, especially in the resonance range. If possible, control movements and vibration movements should not be in the same direction. Finally, under practical working conditions one must always take into consideration the fact that, in addition to vibration stress, further stress factors can also influence performance.

The subjective perception of vibration intensity is an important criterion for evaluation, as is the case for noise. Thus, the subjective vibration perception stated in a multitude of experimental results demonstrates a good basis for the evaluation of vibration in terms of the influence of body posture, vibration frequency, and acceleration intensity, when used in conjunction with knowledge of biodynamic vibration behavior, changes in physiological reactions, and chronic effects.

Chronic changes arising from whole-body vibration occur in the spinal column and the stomach. Since these changes are not specific, however, their causal relationship with vibration stress is difficult to prove. An examination of the many epidemiological studies leads inevitably to the conclusion that long-term vibration stress, which exceeds the "exposure" curve (in accordance with

VDI Guideline 2057 or international standard 2631), leads to an increase in the risk of danger to health, particularly of the spinal column.

Any preventive measure taken by individuals for their personal protection, by technological means, by work organization, or by occupational medicine — are therefore all important.

11. References

Adey WR et al. (1963) EEG in simulated stresses of space flight with special reference to problems of vibration. Electroencephalogr Clin Neurophysiol 15:305–320

Allen GR (1976) Progress on a specification for human tolerance to repeated shock. U.K. Information group on Human Response to Vibration. (unpublished)

Allen RW, Jex HR, Magdaleno RE (1973) Manual control performance and dynamic response during sinusoidal vibration. Aerospace Medical Research Laboratory, Ohio, technical report, pp 73–78

Andreeva-Galanina EC (1957) La maladie vibratoires, ses etilogie, pathogenie et prophylaxie. Proceedings of the 12th international congress of occupational medicine, Helsinki, vol 3, pp 385–387

Andreeva-Galanina EC et al. (1961) Die Vibrationskrankheit (in Russian). Medicina, Leningrad

Andreeva-Galanina EC (1967) Arbeitshygiene bei der Arbeit mit mechanisierten Werkzeugen. Leningrad, pp 1–220

Arnautova-Bulat St (1979) Über die mögliche Wirkung von Vibrationen auf die Wirbelsäule von Eisenbahnern. Arh Hig Rada Toksikol 30:259–266

Ashley C (1970) Equal annoyance contours for the effect of sinusoidal vibration on man. Shock Vibr Bull 41:2

Ashley C (1978) The sensitivity of supine stretcher-borne human subjects to vibrations in three translational and two rotational modes. MIRA Project 5/30/008, Nuneaton 1–15

Bader O (1967) Untersuchung zur Frage der Verwertbarkeit von Kreislaufuntersuchungen als Indikator für die biologische Wirkung mechanischer Schwingungen bei Schlepperfahrern im Arbeitsversuch. Dissertation, University of Bonn, pp 1–79

Bakke S (1931) Röntgenologische Beobachtungen über die Bewegungen der Wirbelsäule. Acta Radiol [Suppl] 13

Bantle IA (1971) Effects of mechanical vibration on the growth and development of mouse embryos. Aerospace Med 42:1087

Barbaso E (1958) Sull'incidenza delle alterazioni della colonna vertebrale nel personale viaggiante di una azienda auto-tramviaria. Med Lavoro 49:630–634

Barnes GR (1979) The effect of aircraft vibration on vision. AGARD – Conference Proceedings, AGARD-CP-267, 12-1–12-11

Bartels P, Dupuis H, Jenik P, Tronich G (1983) Verminderung von Lärm und Vibrationen von Gabelstaplern. Research report for HA 83-010 BMFT, Karlsruhe, pp 1–136

Barton IC (1981) Off-road machine operator vibration measurement methods. SAE Paper 810695:1–19

Bastek R et al. (1977) Comparison of the effects of sinusoidal and stochastic octave-band-wide vibrations – a multi-disciplinary study, parts 1, 2, 3. Int Arch Occup Environ Health 39:143–179

Beck A (1973) Radiologische Beurteilung der Wirbelsäule aus fliegerärztlicher Sicht. Wehrmed Monatsschr 267–276

Beck A (1981) Spätreaktionen der Wirbelsäule bei Flugzeugbesatzungen. Wehrmed Monatsschr 44–50

Békésy von G (1939) Über die Empfindlichkeit des stehenden und sitzenden Menschen gegenüber sinusförmigen Erschütterungen. Akust Z 4:360

Benkendorf L (1953) Über die Behandlung der Seekrankheit. Dtsch Med Wochenschr 78 (12):393-395

Bennett MD et al. (1976) An investigation into some of the human responses to sinusoidal and random vibration in the up-right and semi-reclined seated posture. Shrivenham

Benson AJ (1972) Effect of angular oscillation in yaw on vision. Aerospace Med Assoc 43:44

Berthoz A (1966) Vibrations de basses frequences subies par l'homme. Dissertation, Paris, pp 1-93

Berthoz A (1969) Protection de l'homme contre les vibrations. Lab Phys Trav, Paris, 1-160

Berthoz A, Wisner A (1968) Striated muscles activity and biomechanical effects in man submitted to low frequency vibrations. Electromyography 1 (8):101-109

Berthoz A et al. (1972) Effet des chocs et des vibrations sur le controle de mouvement. Lab Phys Trav Ergon Rapp 30:1-43

Best SG (1945) Propeller balance problems. SAE 53:648

Bez U (1980) Beitrag zur Konzeption von Verkehrsrettungsmitteln. Dissertation D 83, Technical University, Berlin, pp 1-246

Blivaiss BB, Foa PP (1964) Effects of whole-body vibrations on plasma and urinary corticosteroid levels in man. Aerospace Medical Research Laboratory, Ohio, technical report 64-53, pp 1-13

Bobbert G (1967) Schwingungsauswirkung auf den Menschen.VDI Ber 113:95-100

Bödecker H (1937) Über den Einfluß des Seeklimas auf die Histamin-Magensaft-Reaktion. Dissertation, Hamburg

Böhler L (1953) Die Technik der Knochenbruchbehandlung. Maudrich, Wien, p 318

Borscevskij IJ et al. (1963) Obscaja vibracija i ee vlijanie na organizm celoveka. Medgiz, Moskau, pp 1-156

Bosch (1976) Kraftfahrtechnisches Taschenbuch. Firma Bosch, Stuttgart

Brand H, Schnauber H (1980) Abbau von Schwingungsbelastungen in der Eisen- und Stahlindustrie. Humanisierung des Arbeitslebens 6:1-248

Brown Th, Hansen RJ, Yorra AJ (1957) Some mechanical tests on the lumbosacral spine with particular reference to the intervertebral discs. J Bone Joint Surg Am 39 (A):1135-1165

Bruner HD (1960) Methods in medical research, measurement of gastrointestinal mobility in man. Year Book, Chicago, pp 220-221

Buckout R (1964) Effects of whole-body vibration on human performance. Human Factors 6:157-163

Buetti-Bäuml C (1954) Funktionelle Röntgendiagnostik der Halswirbelsäule. Thieme, Stuttgart

Butkovskaja ZM, Kadyskina EN (1970) Sravnitel'naja ocenka vlijanija na organizm celoveka obscej vibracii sinusoidal'noj i slucajnoj formy. gig, Truda Prof Zabol 36-38

Christ E (1973) Beanspruchung und Leistungsfähigkeit des Menschen bei unterbrochener und Langzeit-Exposition mit stochastischen Schwingungen. Dissertation, Technical University, Darmstadt, VDI Ber 11 (17):1-85

Christ W (1963) Aufbaustörungen der Wirbelsäule bei den in der Landwirtschaft tätigen Jugendlichen im Hinblick auf das Schlepperfahren. Grundl Landt 13-15

Christ W, Dupuis H (1963) Der Einfluß vertikaler Schwingungen auf Wirbelsäule und Magen (Röntgenkinematographische Studien). Zentralbl Arbeitsmed Arbeitsschutz Prophyl Ergonomie 13:4-9

Christ W, Dupuis H (1966) Über die Beanspruchung der Wirbelsäule unter dem Einfluß sinusförmiger und stochastischer Schwingungen. Int Z Angew Physiol Arbeitsphysiol 22:258-278

Christ W, Dupuis (1968) Untersuchung der Möglichkeit von gesundheitlichen Schädigungen im Bereich der Wirbelsäule. Med Welt 19:1919-1920, 1967-1972

Coermann R (1938/1939) Untersuchung über die Einwirkung von Schwingungen auf den menschlichen Organismus. Jahrb Dtsch Luftfahrtforsch III 111-III 142 (1938), Dissertation, Technical University, Berlin, pp 73-117

Coermann R (1961) The mechanical impedance of the human body in sitting and standing position at low frequencies. ASD Technical Report 61-492, Ohio, 1-39

Coermann R (1962) Die Wirkung mechanischer Schwingungen auf den Menschen und seine Arbeitsleistung. Werkstattstechnik 52:18-25

Coermann R (1963) The mechanical impedance of the human body in sitting and standing position at low frequencies. In: Lippert (ed) Vibration research. Pergamon

Coermann R (1964) Comparison of the dynamic characteristics of dummies, animals and man. Aerospace Medical Research Laboratory, Ohio

Coermann R (1965) Physiologische Schwingungsprobleme in Fahrzeugen. Zentralbl Verk Med Verk Psych Luft Raumf Med 11:3

Coermann R et al. (1960) The passive dynamic mechanical properties of the human thorax-abdomen system and of the whole-body system. Aerospace Med 31:443–455

Coermann R, Magid EB, Lange KO (1962) Human performance under vibrational stress. Hum Factors, 315–324

Coermann R, Okada A (1964) Übertragung von Erschütterungen auf den Menschen bei verschiedenen Anstellwinkeln der Rückenlehne. Int Z Angew Physiol Arbeitsphysiol 20:398–411

Coermann R, Okada A, Frieling J (1965) Vegetative Reaktionen des Menschen bei niederfrequenter Schwingungsbelastung. Int Z Angew Physiol Arbeitsphysiol 21:150–168

Cohen HC, Wasserman DE, Hornung RW (1977) Human performance and transmissibility under sinusoidal and mixed vertical vibration. Ergonomics 20 (3):207–216

Constant H (1932) Aircraft vibration. J R Aero Soc 36:205

Cope F, Polis BD (1959) Increased plasma glutamicoxaloacetic transaminase activity in monkeys due to nonspecific stress effect. Aerospace Med 30:90

Cottet I, Wisner A, Berthoz A (1968) Application des vibrations mechaniques au traitement expulsif des calculs urinaires. Bull Acad Natl Med 152 (7–8):111–119

Cremona E (1972) Die Wirbelsäule bei den Schwerarbeitern der Eisen- und Stahlindustrie sowie des Bergbaus. Kommiss. Europ. Gem. Generaldir. Soz. Angelegenheiten Dok. Nr 1911/72

Cursiter MC, Harding RH (1974) Electromyographic recordings of shoulder and neck muscles of seated subjects exposed to vertical vibrations. Proc Physiol Soc I Physiol 239:117–118

De Gail P, Lance IW, Neilson PD (1966) Differential effects on tonic and phasic reflex mechanism by vibration in man. J Neurol Neurosurg Psychiatry 29:1–11

Dempsey TK, Leatherwood ID, Clevenson SA (1979) Single axis vibration discomfort criteria. NASA-TP 1422

Dennis JP (1963) A survey of experimental data on the effects of whole-body vibration and of the visual object upon visual performance. Occup Psychol 37:277–282

Dennis JP (1965) The effect of whole-body vibration on a visual performance task. Ergonomics 8 (2):193–205

Dieckmann D (1956) Die Einwirkung mechanischer Schwingungen bis 100 Hz auf den Menschen. Ultraschall Med Grenzgebiete 9 (3):1–10

Dieckmann D (1957) Einfluß vertikaler mechanischer Schwingungen auf den Menschen. Int Z Angew Physiol Arbeitsphysiol 16:519–564

Dieckmann D (1958a) Mechanische Modelle für den vertikal schwingenden menschlichen Körper. Int Z Angew Physiol Arbeitsphysiol 17:67–82

Dieckmann D (1958b) Einfluß horizontaler mechanischer Schwingungen auf den Menschen. Int Z Angew Physiol Arbeitsphysiol 17:83–100

Dieckmann D (1961) Einwirkung mechanischer Schwingungen auf den Menschen. In: Arbeitsphysiologie. Urban und Schwarzenberg, pp 701–717 (Handbuch der gesamten Arbeitsmedizin, vol 1)

Dieckmann D (1963) Über die Einwirkung mechanischer Schwingungen auf den Menschen. Arbeit Leistung 17:12

Dielmann E (1983) Experimentelle Untersuchung zur Belastung der Wirbelsäule durch mechanische Schwingungen bei verschiedenen Fortbewegungsarten. Dissertation, University of Mainz

DIN 1311 (1974) Schwingungslehre, paper 1: Kinematische Begriffe (1974), paper 2: Einfache Schwinger

DIN 3831 (in press) Schutzmaßnahmen gegen die Einwirkung mechanischer Schwingungen auf den Menschen (in preparation)

DIN 4150 (1939) Erschütterungsschutz im Bauwesen

DIN 4150 (1975) Erschütterungen im Bauwesen, paper 2: Einwirkungen auf den Menschen in Gebäuden

DIN 58 220, Blatt 1 (1974) Sehschärfebestimmung, Sehzeichen und Darbietungsbedingungen

Ditchburn RW (1973) Eye-movements and visual perception. Clarendon, Oxford, pp 36 – 62

Doden W (1976) Nystagmus, Nystagmographie. In: Straub (ed) Die ophthalmologischen Untersuchungsmethoden. Enke, Stuttgart, p 2, p 200

Donati P et al. (1982) The subjective equivalence of sinusoidal and random whole-body vibration in sitting position. (Unpublished manuscript)

Dowd PJ (1965) Resistance to motion sickness through repeated exposure to coriolis stimulation. Aerospace Med 452 – 455

Draeger J, Dupuis H (1975) Mechanische Faktoren bei der Auslösung der Amotio retinae. Klin Monatsbl Augenheilkd 166 (4):431 – 435

Drazin DH, Guignard JC (1959) Some effects of low frequency vibration on vision. Sci Memo 21:339 – 342

Dressler (1969) Studie über den Einfluß mechanischer Schwingungen auf das Herz-Kreislauf-System. German institute for occupational medicine, Berlin, GDR

Dudek PE, Ayoub MM, El-Nawawi MA (1973) Optimal work–rest schedules under prolonged vibration. Ergonomics 16 (4):469 – 479

Duffner CR, Hamilton CH, Schmitz MA (1962) Effect of whole body sinusoidal vibration on respiration in human subjects. J Appl Physiol 17:913 – 916

Dupuis H (1960) Schwingungsuntersuchungen bei Schleppern auf einem Rollenprüfstand. Landt Forsch 10 (6):145 – 156

Dupuis H (1966) Arbeitsmedizinische Untersuchungen der Schwingungseinwirkung auf Wirbelsäule und Magen bei Kraftfahrern. 11th international congress of automotive engineering, FISITA, Munich, C 5, 1 – 12

Dupuis H (1969) Zur physiologischen Beanspruchung des Menschen durch mechanische Schwingungen. VDI Ber 11 (7):1 – 168

Dupuis H (1975a) Begriffe und Meßgrößen für mechanische Schwingungen. In: Handbuch der Ergonomie. Zuerl, Steinebach A-9.4.1, pp 1 – 3

Dupuis H (1975b) Physiologische Wirkungen mechanischer Schwingungen. In: Handbuch der Ergonomie. Hanser, München, A-9.4.2, pp 1 – 7

Dupuis H (1976) Belastung und Beanspruchung der Wirbelsäule durch Vibration. In: Die Wirbelsäule in der Flugmedizin. Hippokrates, Stuttgart, pp 48 – 53 (Die Wirbelsäule in Forschung und Praxis, vol 68)

Dupuis H (1977) Human response to vibration under different conditions. Proceedings of the European symposium on life science research in space, Cologne. ESA SP 130:327 – 333

Dupuis H (1979) Lärm und andere physikalische Einflußfaktoren. Z Arb Wiss 33 (1):23 – 26

Dupuis H (1980a) Stand der arbeitsmedizinischen Bewertung mechanischer Schwingungen am Arbeitsplatz. Arb Med Soz Med Präv Med 15 (10):236 – 243

Dupuis H (1980b) Einwirkung berufsbedingter Vibrationen auf die Wirbelsäule. Hippokrates, Stuttgart, pp 45 – 50 (Die Wirbelsäule in Forschung und Praxis, vol 92)

Dupuis H (1981a) Schwingungsmessungen. In: Meßtechnisches Taschenbuch für den Betriebspraktiker – Schall und Schwingungen am Arbeitsplatz. IfaA, Bachem, pp 137 – 177

Dupuis H (1981b) Mechanische Schwingungen und Stöße. In: Schmidtke H (ed) Lehrbuch der Ergonomie, 2nd edn. Hanser, München, pp 222 – 235

Dupuis H (1981c) Stoßbewertung. In: Schmidtke H (ed) Lehrbuch der Ergonomie. Hanser, München, pp 234 – 235

Dupuis H (1981d) Untersuchungen zur Beeinflussung der visuellen Wahrnehmung durch Vibration. Zbl Arb Med 31 (3):90 – 95

Dupuis H (1982) Wirkung mechanischer Schwingungen auf das Hand-Arm-System. Research report no 308 BAU, Wirtschaftsverlag, Bremerhaven, pp 1 – 172

Dupuis H, Preuschen R, Schulte B (1955) Zweckmäßige Gestaltung des Schlepperführerstandes. Landarb Techn 20:1 – 177

Dupuis H, Broicher H-A (1966) Servo-hydraulischer Schwingtisch zur Simulierung von Fahrzeugschwingungen und von stochastischen Abläufen für arbeitsmedizinische Probleme. Automobilt Z 68:41 – 44

Dupuis H, Christ W (1966a) Über das Schwingverhalten des Magens unter dem Einfluß sinusförmiger und stochastischer Schwingungen. Int Z Angew Physiol Arbeitsphysiol 22:149 – 166

Dupuis H, Christ W (1966 b) Untersuchung der Möglichkeit von Gesundheitsschädigungen im Bereich der Wirbelsäule bei Schlepperfahrern. Research report, Max-Planck-Institut für Landarbeit und Landtechnik, Bad Kreuznach

Dupuis H, Christ W (1967 a) Fahrzeugschwingungen und der Einfluß auf Wirbelsäule und Magen. Therapie Nervensystem 7:276−279

Dupuis H, Christ W (1967 b) Über das Verhalten einiger Körperteile bei Schwingungseinwirkung. VDI Ber 113:105−108

Dupuis H, Christ W (1972) Untersuchung der Möglichkeit von Gesundheitsschädigungen im Bereich der Wirbelsäule bei Schlepperfahrern. − Zweite Folgeuntersuchung −. Max-Planck- Institut für Landarbeit und Landtechnik, H. A 72/2, Bad Kreuznach

Dupuis H, Hartung E (1966) Schleppersitzuntersuchungen mit Hilfe eines servo-hydraulischen Schwingungssimulators. Landtechn Forsch 5:164−167

Dupuis H, Hartung E (1972) Arbeitsphysiologische Untersuchungen zur Belastungen von Fahrern auf Rad- und Kettenfahrzeugen durch mechanische Schwingungen. Research report Wehrmedizin BMVg-FBWM 72-2, 1−93

Dupuis H, Hartung E, Louda L (1972) Vergleich regelloser Schwingungen eines begrenzten Frequenzbereiches mit sinusförmigen Schwingungen hinsichtlich der Einwirkung auf den Menschen. Ergonomics 15 (3):237−265

Dupuis H, Hartung E (1973) Arbeitsphysiologische Untersuchungen zur Belastung von Fahrern auf Rad- und Kettenfahrzeugen durch mechanische Schwingungen. Research report Wehrmedizin BMVg-FBWM 73-6, 1−47

Dupuis H, Kleinhanss G (1973) Schockversuche mit Testpuppen und ihre Beurteilung. VDI Ber 210:29−32

Dupuis H et al. (1974) Über den Einfluß stochastischer mechanischer Schwingungen auf physiologische und psychologische Funktionen sowie auf die subjektive Wahrnehmung. Wehrmed Monatsschr 18 (7):193−204

Dupuis H, Draeger J, Hartung E (1975) Vibration transmission to different parts of the body by various locomotions. Biomechanics V-A, 1 A University Park Press, pp 537−543

Dupuis H, Hartung E (1978) The biomechanical behaviour of the bulbus under vibration stress. 1. International conference of mechanics in medicine and biology. Witzstrock, Baden-Baden, pp 76−79

Dupuis H, Hartung E (1979) Tierexperimente zur Ermittlung des biomechanischen Schwingungsverhaltens des Bulbus. Graefes Arch Klin Exp Ophthal 210 (167):167−174

Dupuis H, Hartung E (1980 a) Einfluß von Vibrationen auf die optische Wahrnehmung. Research report Wehrmedizin BMVg-FBWM 80-10, 1−141

Dupuis H, Hartung E (1980 b) Ermittlung des biomechanischen Schwingungsverhaltens menschlicher Bulbi mit Video-Technik. Graefes Arch Klin Exp Ophthal 213:245−250

Dupuis H, Hartung E (1981) Zur Beeinflussung der Schwingungswahrnehmung beim Kranken- und Verletzten-Transport. Notfallmedizin 7:81−90

Dupuis H, Hartung E (1982) Belastung durch Ganz-Körper-Schwingungen in Kraftfahrzeugen und Arbeitsmaschinen. Report TBG, Mainz, pp 1−43

Dupuis H et al. (1982 a) Akute Veränderungen der peripheren Fingerdurchblutung unter Lärm, statischer Belastung und Schwingungsbelastung. Z Arb Wiss 36 (4):243−246

Dupuis H et al. (1982 b) Schwingungsarme Fahrersitze für Nutzfahrzeuge und Arbeitsmaschinen. Humanisierung des Arbeitslebens, BMFT 29, VDI Ber 1−163

Edwards R, Lange KO (1964) A mechanical impedance investigation of human response to vibration. Biophysics Lab WPAFB, Ohio 1−28

Evans GE, Lissner HR (1959) Biomechanical studies on the lumbar spine and pelvis. J Bone Joint Surg 41-A:278−290

EWG 78/764, EWG 83/190 (1978, 1983) EG-Richtlinien zur Angleichung der Rechtsvorschriften der Mitgliedstaaten über den Führersitz von land- und forstwirtschaftlichen Zugmaschinen auf Rädern

Fairbanks AE (1964) Deformation of the human body due to forced vibration. Kentucky Eng 64:8−14

Farfan HF (1979) Biomechanik der Lendenwirbelsäule. Hippokrates, Stuttgart, pp 1−264 (Die Wirbelsäule in Forschung und Praxis, vol 80)

Fassbender HG (1979) Tierexperimentelle licht- und elektronenoptische Untersuchungen über die Entstehung und den Charakter von Vibrationsschäden. Unpublished research report, Mainz

Fischer V et al. (1980) Vibrationsbedingte Wirbelsäulenschäden bei Hubschrauberpiloten. Arb Med Soz Med Präv Med 7:161−163

Fishbein WI, Salter LC (1950) The relationship between truck and tractor driving and disorders of the spine and supporting structures. Indust Med Surg 19:444−445

Fothergill LC (1972) A study of the subjective response to whole-body vibration. Thesis, University of Southampton

Fothergill LC (1973) Psychophysical scaling of human response to whole-body vertical vibration. U.K. Informal group meeting on human response to vibration. (Unpublished)

Fothergill LC, Griffin MJ (1977a) The evaluations of discomfort produced by multiple frequency whole-body vibrations. Ergonomics 20 (3):263−276

Fothergill LC, Griffin MJ (1977b) The subjective magnitude of whole-body vibration. Ergonomics 20 (5):521−533

Franke W (1978) Ärztliche Untersuchungen der Fahrer schwerer Fahrzeuge im Salzbergbau. Kali Steinsalz:253−255

Fraser TM, Hoover GN, Ashe WF (1961) Tracking performance during low frequency vibration. Aerospace, Med 32, pp 829−835

Frenking H (1980) Schwingungsbelastungen auf Baumaschinen. Research report 250, BAU, Wirtschaftsverlag, Bremerhaven, pp 1−199

Freund JL, Dupuis H (1974) Physiologische und pathophysiologische Veränderungen durch Ganz-Körper-Schwingungen. Arb Med Soz Med Präv Med 9 (11):234−236

Freytag I, Puzyna C (1953) Untersuchungen über den Einfluß von Erschütterungen auf den Gesundheitszustand der Schlepperfahrer. A. Univ. M. Curie, Shlodowska VIII/20:335−353

Gaeuman JV, Hoover GN, Ashe WF (1962) Oxygen consumption during human vibration exposure. Aerospace Med 469−474

Garbe C (1981) Gesundheitszustand und gesundheitliche Risiken von Linienbusfahrern in Berlin (West). Reimer, Berlin (Schriftenreihe des Institutes für Sozialmedizin und Epidemiologie des Bundesgesundheitsamtes)

Gauthier GM et al. (1981) Effects of whole-body vibrations on sensory motor system performance in man. Aviat Space Environ Med 473−479

Geller W (1940) Die Seekrankheit und ihre Behandlung. Klin Wochenschr 1310

Gemne G, Taylor W (1984) Hand-arm vibration and the central nervous system. J Low Frequency Noise Vibr (Spec Issue), vol 1

Gierke HE von (1968) Response of the body to mechanical forces − an overview. Ann NY Acad Sci 152 (1):172−186

Gierke HE von (1971) Biodynamic models and their applications. J Acoust Soc Am 50 (6):1397−1413

Glaser EM (1953) Entstehung und Behandlung der Seekrankheit. Dtsch Med Wochenschr 78 (12):392−393

Goethe H (1954) Erkrankungen, Begleiterscheinungen und Folgezustände bei Seekrankheit. Z Tropenmed Parasit 5 (2):232−237

Goethe H (1956) Gedanken zur experimentellen und praktischen Prüfung von Seekrankheitmitteln. Ärztl Praxis 8 (3):8−9

Goethe H (1967) Kinetosen. Kurz Gut 5 (1):12−14

Goethe H (1973) Die Seekinetose. Eine Übersicht über eigene Untersuchungen mit kritischer Betrachtung der Literatur. Hamburg, pp 1−129

Goethe H, Fischer G (1957) Überlegungen zur Erfassung der physikalischen Größen, die zur Entstehung der Seekinetose führen. Zentralbl Verkehrs Med 3 (3):148−153

Goldman DE (1948a) Effect of mechanical vibration on the patellar reflex of the cat. Am J Physiol 155:78

Goldman DE (1948b) A review of subjective responses to vibratory motion of the human body in the frequency range 1−70 c.p.s. Nav. Research Institute, NM 004-001 Rep. 1

Goldman DE, Gierke HE von (1960) The effects of shock and vibration on man. Nav Medical Research Institute, Bethesda, Lect Rev Ser 60 (3):153−198

Goto D, Kanda H (1977) Motion sickness incidence in the actual environment. Hope International JSME Symposium, pp 367–374

Gray R et al. (1976) The effects of 3 hours of vertical vibration at 5 Hz on the performance of some tasks. Royal Aircraft Establ technical report 76011, pp 1–42

Grether WF (1971) Vibration and human performance. Hum Factors 13 (3):203–216

Griffin MJ (1975a) Vertical vibration of seated subjects: effect of posture, vibration level and frequency. Aviat Space Environ Med 46 (3):269–276

Griffin MJ (1975b) Levels of whole-body vibration affecting human vision. Aviat Space Environ Med 46:1033–1040

Griffin MJ (1976a) Eye motion during whole-body vertical vibration. Human Factors 18 (6):601–606

Griffin MJ (1976b) Subjective equivalence of sinusoidal and random whole-body vibration. J Acoust Soc Am 60 (5):1140–1145

Griffin MJ, Whitam EM (1977) Assessing the discomfort of dual-axis whole-body vibration. J Sound Vibr 54 (1):107–116

Griffin MJ et al. (1978) The biodynamic response of the human body and its application to standards. AGARD, Paris A 28-1–A 28-16

Griffin MJ, Lewis CH (1978) A review of the effects of vibration on visual acuity and continuous manual control, part I: Visual acuity. J Sound Vibr 56 (3):383–413

Griffin MJ, Whitham EM (1978) Individual variability and its effect on subjective and biodynamic response to whole-body vibration. J Sound Vibr 58 (2):239–250

Griffin MJ, Whitham EM (1980a) Discomfort produced by impulsive whole-body vibration. J Acoust Soc Am 5:1277–1284

Griffin MJ, Whitham EM (1980b) Time dependency of whole-body vibration discomfort. J Acoust Soc Am. 68 (5):1522–1523

Gruber GJ (1976) Relationship between whole-body vibration and morbidity patterns among interstate truck drivers. NIOSH Report, pp 77–167

Gruber GJ, Ziperman HH (1974) Relationship between whole-body vibration and morbidity patterns among motor coach operators. US DHEW (NIOSH) Contr No HSM-00-72-047

Guignard JC (1960) Physiological effects of mechanical vibration. Proc R Soc Med 53:92–96

Guignard JC (1965) Vibration. In: Gillies (ed) A textbook of aviation physiology. Pergamon, New York, pp 813–894

Guignard JC, Travers PR (1959) Effect of vibrations of the head and of the whole-body on the electromyography activity of postural muscles in man. FPRC/Memo 120, Inst. Aviation Med Farnborough

Guignard JC, Irving A (1960) Effects of low frequency vibration on man. Engineering 190:364–367

Haack M (1953) Über die günstigste Gestaltung der Schleppersitzfederung bei luftbereiften Ackerschleppern mit starrer Hinterachse. Landt Forsch 3 (1):1–13

Häublein HG (1977) Die Gesundheitsrelevanz der körperlichen Schwerarbeit im Bauwesen – eine epidemiologische Studie. Research report of the institute of occupational medicine, Berlin, GDR, pp 1–78

Häupl K (1955) Kieferorthopädischer Gewebeumbau und Muskelreiz. Fortschr Kieferorthop 16:52

Hansson JE, Wikström BO (1981) Comparison of some technical methods for the evaluation of whole-body vibration. Ergonomics 24 (12):953–963

Harris CM, Crede CE (1962) Shock and vibration handbook. McGraw-Hill, New York, 1-5–1-6

Harris CS, Shoenberger RW (1966) Effects of frequency on vibration of human performance. J Eng Psych 5 (1):1–15

Hartung E (1983) Untersuchung zur Beeinflussung der visuellen Leistung des Menschen unter Einwirkung mechanischer Schwingungen. Dissertation, University of Bremen

Hawel W (1969) Untersuchung zur Wirkung vertikalen Schwingens auf die Stimmung von Versuchspersonen. Zentralb Verkehr Med 15:83–89

Heide R (1978) Zur Wirkung langzeitiger beruflicher Ganzkörpervibrationsexposition. Dissertation, University of Berlin, GDR

Heide R, Seidel H (1978) Folgen langzeitiger beruflicher Ganzkörpervibrationsexposition. Z Ges Hyg 24:153–159

Helberg W, Sperling E (1941) Verfahren zur Beurteilung der Laufeigenschaften von Eisenbahnwagen. Org Fortschr Eisenbahnw 96:12

Helling J (1964) Elektrisches Modellverfahren zur Untersuchung des Federungsverhaltens von Sattelkraftfahrzeugen. D Kraftfahrtf Straßenverk T 171:1−35

Helling J (1978) Einwirkung mechanischer Schwingungen auf den liegenden Menschen. Unpublished research report IKA, Aachen, 1−75

Hilfert R et al. (1981) Probleme der Ganzkörperschwingungsbelastung von Erdbaumaschinenführern. Zentralbl Arb Med 31:4−5, part 1:152−155, part 2:199−206

Hirsch C (1955) The reaction of intervertebral discs to compression forces. J Bone Joint Surg 37 (A):1189−1196

Hochberg K, Nöller HG, Kelly T (1964) Magenuntersuchungen mit der Heidelberger Kapsel. Münch Med Wochenschr 106 (17):789−798

Hoffmann H et al. (1969a) Untersuchung zur Frage der Verwendbarkeit von Kreislaufuntersuchungen als Indikator für die biologische Wirkung mechanischer Schwingungen bei Schlepperfahrern im Arbeitsversuch. Zentralbl Verkehr Med Verkehr Psych Luft Raumf Med 15 (1):35−49

Hoffmann H et al. (1969b) Versuche zur Frage der Verwendbarkeit des Kreislaufverhaltens als Indikator für die biologische Wirkung mechanischer Schwingungen beim Transport von Personen im Krankenwagen. Zentralbl Verkehr Med Verkehr Psych Luft Raumf Med 15:27−34

Hoffmann H et al. (1970) Kreislaufuntersuchungen bei Kraftfahrzeugführern unter variierten Fahrbedingungen. Zentralbl Verkehr Med 16:3−4, 192−233

Holland CL (1967) Performance effects of long term random vertical vibration. Hum Factors 9 (2):93−104

Holloway RB, Brumaghim SH (1972) Tests and analysis applicable to passenger ride quality of large transport aircraft. NASA TM X-2620, 91−113

Hood WB et al. (1965) Circulatory and respiratory effects of whole-body vibration in anaesthetized dogs. J Appl Physiol 20:1157−1162

Hood WB et al. (1966) Cardiopulmonary effects of whole-body vibration in man. J Appl Physiol 21:1725−1731

Hornick RJ (1962) The effects of whole-body vibration in the three directions upon human performance. J Eng Psych 1:93−101

Hornick RJ, Boettcher CA, Simons AK (1961) The effect of low frequency, high amplitude, whole-body, longitudinal and transverse vibration upon human performance. Bostrom Research Laboratory, pp 1−54

Horst M (1982) Mechanische Beanspruchung der Wirbelkörperdeckplatte. Hippokrates, Stuttgart, pp 1−88 (Die Wirbelsäule in Forschung und Praxis, vol 95)

Huang BK, Suggs CW (1965) Vibration studies of tractor operators. Am Soc Agr Eng Med Pap No 65-610; 1−10

Ilgmann W (1979) Ergonomische Untersuchung über die Einwirkung rotatorischer Schwingungen, Beanspruchung durch Rollschwingungsbelastung. BMVg-FBWT 79-33, 1−122

ILO (1977) Übereinkommen (Nr. 148) über den Schutz der Arbeitnehmer gegen Berufsgefahren infolge von Luftverunreinigung, Lärm und Vibrationen an den Arbeitsplätzen. Genf

ILO (1980) Änderung der Liste der Berufskrankheiten im Übereinkommen (No 121) über Leistungen bei Arbeitsunfällen und Berufskrankheiten. Int Arbeitskonferenz, 66. Meeting, Genf

Irwin AW (1981) Study and evaluation of human response to pure and combined forms of low frequency motion at various levels. International workshop on research methods in human motion and vibration studies. New Orleans

ISO 2631 (1978) Guide for the evaluation of human exposure to whole-body vibration, 2nd edn

ISO 2631, Addendum 2 (1982) Evaluation of exposure to whole-body z-axis vertical vibration in the frequency range 0.1 to 0.63 Hz

ISO 2631, Amendment 1 (1982) Guide for the evaluation of human exposure to whole-body vibration

ISO 5982 (1981) Vibration and shock − mechanical driving point impedance of the human body

ISO 7096 (1982) Earth-moving machinery – Operator seat – Transmitted vibration

Ivanovitsch E, Antov G, Kazakova B (1981) Liver changes under combined effect of working environmental factors. Int Arch Environ Health 48:41–47

Jacklin HM (1936) Human reaction to vibration. SAE J 39:401–407

Janeway RN (1949) Passenger vibration limits. SAE J 56:48

Jankovich JP (1971) Structural development of bone in the rat under earth gravity, simulated weightlessness, hypergravity and mechanical vibration. NASA CR-1823

Jankovich JP (1972) The effects of mechanical vibration on bone development in the rat. J Biomechan 5:241–250

Jansen G (1966) Psychosomatische Wirkungen des Lärms. Mitt MPG Förd Wiss 5:293–309

Johnson WH (1954) Head movements and motion sickness. International record of medicine & GP Clinics, 638–640

Jones AJ, Saunders DJ (1974) A scale of human reaction to whole-body vertical sinusoidal vibration. J Sound Vibr 35:503–520

Junghanns H (1931) Altersveränderungen der menschlichen Wirbelsäule. 3. Häufigkeit und anatomisches Bild der Spondylosis deformans. Arch Klin Chir 166:120

Junghanns H (1979) Die Wirbelsäule in der Arbeitsmedizin. Hippokrates, Stuttgart, vol 1, pp 103–127; vol 2, pp 132–138, 185–200

Kanda H, Goto D, Tanabe Y (1977) Ultra-low frequency ship vibrations and motion sickness incidence. Ind Health 15 (1):1–12

Kazarian L (1972) Dynamic response characteristic of the human vertebral column. Acta Orthopaed Scand [Suppl] 146

Kazarian L, Graves GA (1977) Compressive strength characteristics of the human vertebral centrum. Spine 2 (1):1–14

Keidel WD (1956) Vibrationsrezeption. Der Erschütterungssinn des Menschen. Universitäts-Bund e.V., Erlangen, pp 1–154 (Erlanger Forschungen, series B, vol 2)

Kersten E (1966) Überlastungsschäden bei Hochseefischern und ihre Beurteilungen im Sinne der VO über Melde- und Entschädigungspflicht bei Berufskrankheiten. Z Ges Hyg 179–182

Kirbey RH et al. (1975) Effects of vibration in combined axes on subjective evaluation of ride quality. NASA TM X-3295, 355–371

Kleinhanss G, Dupuis H (1971) Über die Beanspruchung des Menschen bei simulierter Druckeinwirkung auf Schutzbauten. Biomed Techn 16 (1):28–32

Klosterkötter W (1974) Änderung der peripheren Durchblutung bei Arbeiten mit Motorsägen. Unpublished research report, Essen

Klotter K (1978) Einfache Schwinger, 3rd edn. In: Klotter K, Benz G (eds) Technische Schwingungslehre, vol 1. Springer, Berlin Heidelberg New York, pp 1–425

Köhl U (1975) Les dangers encourus par les conducteurs des tracteurs. Arch Mal Prof 36:145–162

Köhne G (1981) Die beim Betrieb von Erdbaumaschinen auftretenden niederfrequenten mechanischen Ganzkörperschwingungen, die ergonomischen und arbeitsmedizinischen Zusammenhänge und die daraus sich ergebenden Folgerungen. Dissertation, RWTH, Aachen, pp 1–190

Köhne G, Zerlett G, Duntze H (1982) Ganzkörperschwingungen auf Erdbaumaschinen. Humanisierung des Arbeitslebens. VDI Ber 32:1–366

Kohlrausch (1976) Cited from Doden (1931) Nystagmus, Nystagmographie. In: Straub (ed) Die ophthalmologischen Untersuchungsmethoden, vol 2. Enke, Stuttgart, p 192

Kozlov VN, Kiseleva NP (1971) Versuch der elektroencephalographischen Untersuchung von Traktoristen während Feldarbeiten. Gig San 36:106–107

Krapac L (1976) Degenerative Veränderungen der Gelenke und der Wirbelsäule als Folge von beruflichen und Umwelteinflüssen. Arh Hig Rada Toxikol 27:233–241

Krämer J (1978) Bandscheibenbedingte Erkrankungen (Ursachen, Diagnose, Behandlung, Vorbeugung, Begutachtung). Thieme, Stuttgart

Krause H (1963) Das schwingungsmechanische Verhalten der Wirbelsäule. Int Z Angew Physiol Arbeitsphysiol 20:125–155

Kristen H, Lukeschitsch G, Ramach W (1981) Untersuchung der Lendenwirbelsäule bei Kleinlasttransportarbeitern. Arb Med Soz Med Präv Med 16:226–229

Kubic S (1966) Gesundheitliche Schäden bei Traktoristen. Proceedings of the international congress on occupational Health, Vienna, pp 375−377

Kummer B (1959) Bauprinzipien des Säugerskelettes. Thieme, Stuttgart

Kunz F, Meyer HR (1969) Rückenbeschwerden und Wirbelsäulenbefunde bei Führern schwerer Baumaschinen. Z Unfallmed Berufskr 62:178−189

Laarmann A (1977) Berufskrankheiten nach mechanischen Einwirkungen, 2nd edn. Enke, Stuttgart

Lamb TW et al. (1966) Nature of vibration hyperventilation. J Appl Physiol 21:404−410

Lange HJ (1981) Problem des Fehlens geeigneter Vergleichkollektive bei epidemiologischen Studien in der Arbeitsmedizin. Arb Med Soz Med Präv Med 112−116

Lange W (1971) Untersuchungen über Wahrnehmungsstärken und übertragene Schwingkräfte bei der vertikalen Einwirkung von periodischen Schwingungsgemischen auf den Menschen. Dissertation, University of Braunschweig, pp 1−158

Lange W (1974a) Subjektive Schwingungswahrnehmung und Bewertung von Ganzkörperschwingungen. Arb Med Soz Med Präv Med 11:240−241

Lange W (1974b) Zur Beurteilung von Schwingungsgemischen, die über die Sitzfläche auf den Menschen einwirken. Eur J Applied Physiol 33:151−170

Lange W, Coermann R (1965) Relativbewegungen benachbarter Wirbel unter Schwingungsbelastung. Int Z Angew Physiol Arbeitsphysiol 21:326−334

Lavault P (1962) Quelque aspects de la pathologie du rachis chez le conducteur de tracteur agricole. Concours Médical 84:5863−5875

Lebedeva AF, Puskin VZ (1950) Physiologische Bewertung und Versuch der Normierung allgemeiner Vibration. Gigiena Sanit 10:3−10

Lee RA, King AJ (1971) Visual vibration response. J Appl Physiol 30 (2):281−286

Lehmann G, Dieckmann D (1956) Die Wirkung mechanischer Schwingungen (0,5−100 Hz) auf den Menschen. Research report Wirtschafts- und Verkehrsministerium NRW, 362

Lewis CH, Griffin MJ (1976) The effects of vibration on manual control performance. Ergonomics 19:203−216

Lewis CH, Griffin MJ (1978) A review on the effects of vibration on visual acuity and continuous manual control, part 2. Continuous manual control. J Sound Vibr 56 (3):415−457

Lewis CH, Griffin MJ (1979a) The effect of character size on the legibility of numeric displays during vertical whole-body vibration. J Sound Vibr 67 (4):562−565

Lewis CH, Griffin MJ (1979b) Mechanism of the effects of vibration frequency, level and duration on continuous manual control performance. Ergonomics 22 (7):855−889

Lewis CH, Griffin MJ (1980) Predicting the effects of vibration frequency and axis and seating conditions on the reading of numeric displays. Ergonomics 23 (5):485−501

Liebeskind D (1970) Berufskrankheiten im Röntgenbild. Barth, Leipzig

Lindner GS (1962) Mechanical vibration effects of human beings. Aerospace Med 33:939−950

Loeb M (1954) A preliminary investigation of the effects of whole-body vibration and noise. Army Med Res Lab Report 145

Loeckle WE (1950) The physiological effects of mechanical vibration. In: German aviation medicine World War 2. Washington, pp 716−723

Loeckle WE (1941) Über die Wirkung von Schwingungen auf das vegetative Nervensystem und die Sehnenreflexe. Luftfahrtmedizin 5:305−316

Louyot P, De Ren G, Jouret (1954) Le rachis des chauffeurs de locomotives. Association de Médicine du Travail du Nord, 22 Mai, Lille

Magid EB, Coermann R (1960) The reaction of the human body to extreme vibrations. Proc Inst Envir Sci 135

Magid EB, Coermann R, Ziegenrücker GH (1960) Human tolerance to whole body sinusoidal vibration. J Aerospace Med 31:915−924

Mallock A (1902) Broad of trade report, App 5. London

Mandel MJ (1963) Short time tolerance and pulse response in man to sinusoidal vibration in the semi supine position in x, y and z axes. Aerospace Med 34 (3):260−261

Mandel MJ, Robinson FR, Luce AE (1962) SGOT-levels in man and in the monkey following physical and emotional exertion. Aerospace Med 33:1216−1233

Martin B et al. (1980) Effects of whole-body vibrations on standing posture in man. Aviat Space Environ Med 778−787

McFarland RA (1953) Human factors in air transport design. McGraw-Hill, New York, p 703

Menshov AA (1962) Zur Frage der Vorbeugung der Lärm- und Vibrationseinwirkung in der modernen automatisierten Produktion. Z Ges Hyg 9:186 – 189

Menshov AA (1967) Stoßartige Vibration – ihre hygienische Bewertung, Normierung und Prophylaxe unter Produktionsbedingungen. Dissertation, University of Kiew

Mertens H (1978) Nonlinear behaviour of sitting humans under increasing gravity. Aviat Space Environ Med 49 (1):287 – 298

Milby TH, Spear RC (1974) Relationship between whole-body vibration morbidity patterns among heavy equipment operations. NIOSH-Publications No 74 – 131

Miwa T (1967) Evaluation methods for vibration effect, part 1. Measurements of threshold and equal sensation contours of whole-body for vertical and horizontal vibrations. Ind Health 5:183 – 205

Miwa T (1968 a) Evaluation methods for vibration effect, part 5. Calculation method of vibration greatness level on compound vibrations. Ind Health 6:11 – 17

Miwa T (1968 b) Evaluation methods for vibration effect, part 6. Measurements of unpleasant and tolerance limit levels for sinusoidal vibrations. Ind Health 6:18 – 27

Miwa T (1968 c) Evaluation methods for vibration effect, part 7. The vibration greatness of the pulses. Ind Health 6:143 – 164

Miwa T (1969) Evaluation methods for vibration effect, part 8. The vibration greatness of random waves. Ind Health 7:89 – 115

Miwa T (1975) Mechanical impedance of human body in various postures. Ind Health 13 (1):1 – 22

Miwa T, Yonekawa Y (1969) Evaluation methods for vibration effect, part 9. Response to sinusoidal vibration at lying posture. Ind Health 7:116 – 126

Money KE (1970) Motion sickness. Physiol Rev 50 (1):1 – 39

Moseley M, Lewis CH, Griffin MJ (1982) Sinusoidal and random whole-body vibration: comparative effects on visual performance. Aviat Space Environ Med 1000 – 1005

Müller EA (1939) Die Wirkung sinusförmiger Vertikalschwingungen auf den sitzenden und stehenden Menschen. Arbeitsphysiologie 10 (5):459 – 476

Müller EA, Franz H (1952) Energieverbrauchsmessungen bei beruflicher Arbeit mit einer verbesserten Respirations-Gasuhr. Arbeitsphysiologie 14:499 – 504

Müller-Limmroth W (1959) Elektrophysiologie des Gesichtssinnes, Theorie und Praxis der Elektroretinographie. Springer, Berlin Göttingen Heidelberg, pp 109 – 123

Müller-Limmroth W (1961) Die Ermüdung des Kraftfahrers physiologisch betrachtet. Bundesverkehrswacht Drucksache 33:3 – 18

Nachemson A, Morris JM (1964) In vivo measurements of intradiscal pressure. J Bone Joint Surg 46-A:1077 – 1092

Nakamura M (1941) Experimental study of effect of vibration upon internal ear. Far East Scient Bull 1:29

Nickerson J, Coermann R (1962) Internal body movements resulting from externally applied sinusoidal forces. Aerospace medical research laboratory, Ohio, technical report, 62 – 81, pp 1 – 16

Nickerson J, Paradijeff A, Feinhandler HS (1963) A study of the effects of externally applied sinusoidal forces on the eye. Aerospace medical research laboratory, Ohio, technical report 64, p 120

Nickerson J, Drazic M (1966) International body movement along three axes resulting from externally applied sinusoidal forces. Aerospace medical research laboratory, Ohio, technical report 66 – 102, pp 1 – 12

Nickolson AN, Guignard JC (1966) Electrocortiogramm during whole-body vibration. Electroencephalogr Clin Neurophysiol 20:494 – 505

Nigg BM, Neukomm PA (1973) Erschütterungsmessungen beim Skifahren. Med Welt Forsch Praxis 24 (48):1883 – 1885

Nixon CW (1962) Inference of selected vibrations upon speech, I. range of 10 cps to 50 cps. J Audit Res 2:247 – 266

Nixon CW, Sommer HC (1963) Inference of selected vibrations upon speech, III. range of 6 cps to 20 cps for semi-supine talkers. Aerospace Med 34 (11):1012 – 1017

NN (1976) Berufskrankheiten-Verordnung (BeKV) – BGBl I, 3329, 8 December

Nöller HG, Khodabakhsh (1964) Die Säurebildungsleistung des Magens und ihre Individualstreuung. Fortschr Med 82 (7):264−268

Oates JC et al. (1978) Pulsating forces in orthodontic treatment. Am J Orthod 74 (5):577−586

Oborne DC, Clarke MJ (1974) The determination of equal comfort zones for whole-body vibration. Ergonomics 17:769−782

Oborne DJ, Boarer P, Heath TO (1981) Variations in response to whole body vibration intensity dependent effects. Ergonomics 24 (7):523−530

O'Hanlon JF, McCauley ME (1974) Motion sickness incidence as a function of the frequency and acceleration of vertical sinusoidal motion. Aerospace Med 45:366−369

Ohlbaum KM (1976) Mechanical resonant frequency of the human eye in vivo. Aerospace medical research laboratory, Ohio, technical report 75, p 113

Okada A et al (1972) Temporary hearing loss induced by noise and vibration. J Acc Soc Am 4 (2):1240−1248

Oshima M (1963) The effect of vibration on the visual acuity. Japan Publ Trading Co, pp 108−111

Parks DL (1962) Defining human reaction to whole-body vibration. Hum Factors 4:305−314

Parsons KC, Griffin MJ (1978a) The effect of rotational vibration in roll and pitch axes on the discomfort of seated subjects. Ergonomics 21 (8):615−625

Parsons KC, Griffin MJ (1978b) The effect of the position of the axis of rotation on the discomfort caused by whole-body roll and pitch vibrations of seated persons. J Sound Vibr 58 (1):127−141

Parsons KC, Whitham EM, Griffin MJ (1979) Six axis vehicle vibration and its effects on comfort. Ergonomics 22 (2):211−225

Pauwels F (1960) Eine neue Theorie über den Einfluß mechanischer Reize auf die Differenzierung der Stützgewebe. Z Anat Entwickl Gesch 121:478

Petrov P (1968) EEG proncvanija pri vibraconna bolest. Transp Med Westi 13 Sofia, 4:6−16

Pfander F (1978) Wird die lärmbedingte Hörgefährdung bei gleichzeitiger Vibration im Infraschallgebiet verstärkt? 49th Annual meeting of the German Society of otorhino laryngology, Hamburg

Pinter I (1975) Gehörschädigung bei Traktorfahrern in der Forstwirtschaft. Internationale Arbeitskonferenz „Ergon. Stand. i. d. Ldw.", Potsdam, Sond-H, E 171/75, II/13/1

Pleszczynski W, Christ E, Dupuis H (1973) Steuerfehler unter Schwingungseinfluß. Automobilt Z 8:281−284

Postlethwaite F (1944) Human susceptibility to vibration. Eng 157:61

Pradko F (1965) Human response to random vibration. Shock Vibr Bull 34 (4):173−190

Puschkina NM (1961) Über bestimmte biochemische Blutindexe bei Arbeitern, die Erschütterungen ausgesetzt sind. Gig Truda Prof Zabol 2:29−32

Ramazzini B (1977) Untersuchung von den Krankheiten der Künstler und Handwerker. Weidmann, Leipzig (1718), reprinted by Zentrales Antiquariat DDR, Leipzig, pp 325−327

Rao BKN, Jones B (1975) Some studies on the measurement of head and shoulder vibration during walking. Ergonomics 18 (5):555−566

Raymond V (1956) Die Wirkung von Erschütterungen auf die Fahrer von schweren Fahrzeugen. Arch Mal Profess Méd Trav Secur Soc 17:301−310

Reiher H, Meister FJ (1931) Die Empfindlichkeit des Menschen gegen Erschütterungen, Forschungen auf dem Gebiet des Ingenieur-Wesens. Z Techn Mech Thermodyn 2 (11):381−386

Reinhardt RF (1959) Motion sickness: a psychophysiologic gastrointestinal reaction? Aerospace Med 802−805

Rentsch HJ (1960) Ulcuskrankheit und Landarbeit. Dissertation, University of Berlin, GDR

Roll JP et al. (1980) Effects of whole-body vibration on spinal reflexes in man. Aviat Space Environ Med 1227−1233

Rosegger R (1966) Über vorzeitige Aufbraucherscheinungen der Wirbelsäule bei Schlepperfahrern und ihre mögliche Abhängigkeit von Erschütterungen. Dissertation, University of Erlangen

Rosegger S (1970) Vorzeitige Aufbraucherscheinungen bei Kraftfahrern. Z Orthop Grenzgeb 108:510−516

Rosegger R, Rosegger S (1960) Arbeitsmedizinische Erkenntnisse beim Schlepperfahren. Arch Landtechn 2:3−65

Roux·X (1894) Funktionelle Anpassung. Encyclopäd Jahrbuch 4:14

Rowlands GF (1977) The transmission of vertical vibration to the heads and shoulders of seated men. Techn Rep 77068, Royal Aircr Establ 1–79

Rubinstein L, Kaplan R (1968) Some effects of y-axis vibration on visual acuity. AMRL-TR-68-19, 1–32

Rublack H (1978) Wirkungen mechanischer Schwingungen auf den Organismus. Z Ges Hyg 24 (9):649–666

Rudat W (1952) Die Bewegungskrankheiten – Ätiologie und Therapie. Dtsch Med J 3 (15/16):345–346

Rühmann H (1978) Untersuchung über den Einfluß der mechanischen Eigenschaften von Bedienelementen auf die Steuerleistung des Menschen bei stochastischen Rollschwingungen. Research report Wehrtechn BMVg-FBWT 78-11, 1–190

Rumyantsew GK, Chumak KI (1966) Knöcherne Veränderungen der Wirbelsäule von Betonarbeitern als Auswirkung einer Ganz-Körper-Vibration hoher Frequenz. Gig Truda 10:6–9

Ruppe K (1971) Literaturstudie über die Ergebnisse experimenteller und epidemiologischer Untersuchungen zur Belastungswirkung mechanischer Schwingungen auf den Menschen. German institute for occupational medicine, Berlin

Sandover J (1978) Modelling human response to vibration. Aviat Space Environ Med 49 (1):335–339

Sandover J (1981) Vibration, posture and low-back disorders of professional driver. Department Human Science University of Technique, Loughborough, Rep No DHS 402:1–141

Sassor H-J, Krause H (1966) Auswirkungen mechanischer Schwingungen auf den Menschen. RKW-Serie Arbeitsphysiologie Arbeitspsychologie, Beuth, Berlin, pp 3–87

Schadewaldt H (1967) Zur Geschichte der Seekrankheit. Med Welt 18:2258–2265

Schäfer N (1977) Untersuchungen zur ergonomischen Beurteilung von Fahrzeugsitzen unter besonderer Berücksichtigung des Schwingungsübertragungsverhaltens. Thesis, Technical University, Darmstadt, pp 1–147

Schäfer N, Dupuis H, Hartung E (1982) Schwingungsminderung am Arbeitsplatz. Research report no 305 BAU, Wirtschaftsverlag, Bremerhaven, pp 1–200

Scheibe W (1979) Beurteilung von Belastung, Aktivität und Beanspruchung des Menschen bei kontinuierlicher und unterbrochener Exposition mit vertikalen Fahrzeugschwingungen in Simulations- und Feldexperimenten. VDI Ber 11 (31):1–233

Schmidt U (1969) Vergleichende Untersuchungen an Schwerlastwagenfahrern und Büroangestellten zur Frage der berufsbedingten Verschleißschäden an der Wirbelsäule und den Gelenken der oberen Extremitäten. Dissertation, Humboldt-University, Berlin, GDR

Schmidt EG (1981) Belastung der Besatzung von Seeschiffen durch mechanische Schwingungen. Research report no 275 BAU, Wirtschaftsverlag, Bremerhaven, pp 1–276

Schmidtke H (1974) Einfluß mechanischer Schwingungen auf visuelle Informationsaufnahme und motorische Koordination. Arb Med Soz Med Präv Med 9 (11):236–239

Schmidtke H (1975) Einfluß stochastischer vertikaler Beschleunigungskräfte auf die Sehschärfe bei freiäugigem Sehen. Z Arb Wiss 29 (1):30–32

Schmitz MA (1959) The effect of low frequency, high amplitude whole body vertical vibration on human performance. Progr Report no 2a, Milwaukee, Bostrom Research Laboratory, pp 1–58

Schmitz MA, Simons A (1959) Man's response to low-frequency vibration. ASME-Publ. 59-A-20, 1–11

Schmitz MA, Boettcher CA (1960) Some physiological effects of low-frequency, high-amplitude vibration. ASME-Publ Paper no 60-PROD-17, 1–5

Schnauber H, Weigelt P (1981) Schwingungen an Steh-Arbeitsplätzen. Research report no 285 BAU, Wirtschaftsverlag, Bremerhaven, pp 1–128

Schober H (1976) Prüfung der Sehfunktion. In: Straub (ed) Die ophthalmologischen Untersuchungsmethoden. Enke, Stuttgart, pp 273–324

Schoknecht G, Barich G (1978) Ist die Häufung von Wirbelsäulenveränderungen bei Berufskraftfahrern erhöht? Arb Med Soz Med Präv Med 13:281–283

Schütz E (1966) Physiologie, 9th, 10th edn. Urban and Schwarzenberg, Munich, pp 251–258

Schulze KJ, Polster J (1979) Berufsbedingte Wirbelsäulenschäden bei Traktoristen und Landwirten. Beitr Orthop Traum 26:356-362

Segel L et al. (1981) State of knowledge review: relationship of truck ride vibration to highway safety. UM-HSRI 81-57 Michigan, 1-84

Seidel B, Tröster FA (1970) Ergebnis einer gezielten Reihenuntersuchung von 60 Traktoristen auf Gesundheitsschäden durch Lärm und Vibration. Z Ges Hyg 447-450

Seidel H (1975) Systematische Darstellung physiologischer Reaktionen auf Ganzkörperschwingungen in vertikaler Richtung (z-Achse) zur Ermittlung von biologischen Bewertungsparametern. Tribüne, Berlin, Ergon Ber 15:18-39

Seidel H et al. (1980) On human response to prolonged repeated whole-body vibrations. Ergonomics 23 (3):191-211

Selye H (1949) The general adaptation syndrome and the disease of adaptation. J Clin Endocrinol 6:117-230

Serbitzer J (1974) Messung und Beurteilung von Schwingungseinwirkungen auf den Menschen. Tribüne, Berlin, Arbeitsschutz 34:1-140

Sergl HG (1974) Zur Theorie des Knochenwachstums und seiner Steuerung. Fortschr Kieferorthop 35:323-331

Sergl HG (1983) Tierexperimentelle Untersuchungen zur „Erschütterungstheorie". Fortschr Kieferorthop 44:28-38

Seris HJ (1969) Les vibrations méchaniques: action sur l'homme, prévention. Rev des Corps de Santé des Amrées Terre, Mer, Air, X, 1

Shapiro E, Roeber FW, Klempner LS (1979) Orthodontic movement using pulsating force-induced piezoelectricity. Am J Orthop 76 (1):59-66

Shoenberger RW (1972) Human response to whole-body vibration. Percept Mot Skills 34:127-160

Shoenberger RW (1974) An investigation of human information processing during whole-body vibration. Aerospace Med 143-153

Shoenberger RW (1975) Subjective response to very low frequency vibration. Aeospace Med Assoc Meet Proc 159-160

Shoenberger RW (1979) Psychophysical assessment of angular vibration: comparison of vertical and roll vibration. Aviat Space Environ Med 7:688-691

Shoenberger RW (1980) Psychophysical comparison of vertical and angular vibrations. Aviat Space Environ Med 8:759-762

Simic D (1970) Beitrag zur Optimierung der Schwingungseigenschaften des Fahrzeuges - Physiologische Grundlagen des Schwingungskomforts. Dissertation, Technical University, Berlin, pp 1-167

Simons AK, Schmitz MA (1958) The effect of low-frequency high amplitude whole-body vibration on human performance. Bostrom Research Laboratory Milwaukee, Progr 1 Report

Sjøflot L, Suggs CW (1973) Human reactions to whole-body transverse angular vibrations compared to linear vertical vibrations. Ergonomics 16 (4):455-468

Snyder FW (1965) Vibration and vision. Hum Factors 183-201

Soliman JI (1968) A scale for the degree of vibration perceptibility and annoyance. Ergonomics 11 (2):101-122

Sommer HC (1973) The combined effects of vibration, noise and exposure duration on auditory temporay threshold shift. Aerospace medical research laboratory, Ohio, technical report 73, p 34

Spangenberg WW (1962) Die Kinetosen. Verkehr Med 9 (5/6):197-209

Spear RC, Keller CA (1976) Morbidity studies of workers exposed to whole body vibration. Arch Environ Health 16:141-145

Spear RC et al. (1976) Morbidity patterns among heavy equipment operators exposed to whole body vibration. NIOSH-Public no 77-120

Spilberg PI (1962) Elektroenzephalographische Untersuchungen bei der Vibrationskrankheit, bedingt durch die Wirkung der allgemeinen Vibration. Gig, Truda 4:14

Splittgerber H (1972) Untersuchungen über die Wahrnehmungsschwelle des Menschen bei einwirkenden mechanischen Schwingungen. Gesundheits-Ing 93 (4):113-118

Stankovic P (1969) Über die Beeinflussung der Darmmotorik durch mechanische Schwingungen. Beitr Klin Chir 217:613-615

Starlinger H, Hawel W, Rutenfranz J (1969) Untersuchungen zur Frage der Catecholaminaus-
    scheidung im Harn als Kriterium für emotionalen Streß unter verschiedenen Um-
    gebungsbedingungen. Int Z Angew Physiol 27:1–14
Stave AM (1979) The influence of low frequency vibration on pilot performance. Ergonomics
    22 (7):823–835
Steinhäuser H, Bolt W (1979) Arbeit und Verkehr. In: Orthopädie in Praxis und Klinik. Thie-
    me, Stuttgart (Allgemeine Orthopädie, vol 1)
Suzuki Y et al. (1959) Studies on the influence of vibration upon the gastric emptying.
    Tokushima Med (6) 4:192–202, 231–242
Szameitat P (1976) Mechanische Schwingungseinwirkung auf den liegenden Menschen. Dis-
    sertation, Technical University, Darmstadt, pp 1–226
Szameitat P, Dupuis H (1976) Über die Beeinflussung des liegenden Menschen durch
    mechanische Schwingungen. Arbeiten MPI für Landarbeit und Landtechnik A-76-1:1–116
Taub HA (1966) Dial-reading performance as a function of frequency of vibration and head
    restraint system. Aerospace medical research laboratory, Ohio, technical report 16,
    pp 1–21
Teare RJ, Snyder FW (1963) Human hearing and speech during whole-body vibration.
    Technical Report 3, Boeing Comp Contr no 2994 (00)
Teare RJ, Parks DL (1963) Visual performance during whole-body vibration. Technical Re-
    port D 3-3512-4, 1–28
Temple WE et al. (1964) Man's short-time tolerance to sinusodial vibration. Aerospace Med
    35 (10):923–930
Tippelmann M (1964) Das Verhalten der eosinophilen Blutzellen unter verschiedenen Be-
    lastungen bei gleichen Individuen. DVL-Ber 310
Usutani S, Ishida R, Ogino Y (1965) Influence of the vibration on the coefficient of digesti-
    bility of food in the rats. Hirosaki University Department of Public Health, pp 407–414
UVV (1976) Unfallverhütungsvorschrift VBG Nr 40: Bagger, Planiergeräte, Schürfgeräte und
    Spezialmaschinen des Erdbaues
VDI 2062 (1976) Schwingungsisolierung, paper 1: Begriffe und Methoden, Jan., paper 2:
    Isolierelemente, Jan.
VDI 2057 (1979/81) Beurteilung der Einwirkung mechanischer Schwingungen auf den
    Menschen, paper 1: Grundlagen, Gliederung, Begriffe, Entw. Febr. (1979), paper 2:
    Schwingungseinwirkung auf den menschlichen Körper, Mai (1981), paper 3: Schwingungs-
    beanspruchung des Menschen, Entw. Feb. (1979)
VDI 2782 (1971) Empfehlungen für die Gestaltung von Fahrzeugführersitzen in Kraft-
    fahrzeugen
VDI 2783 (1972) Empfehlungen für die Gestaltung von Fahrgast- und Beifahrersitzen
Vogt LH (1968) Mechanical impedance of the sitting human under sustained acceleration.
    Aerospace Med 39 (7):665–679
Vogt LH (1973) Mechanical impedance of supine humans under sustained acceleration.
    Aerospace Med 44 (2):123–128
Vogt LH, Krause HE (1973) Schwingungsmessungen an vier Körpersegmenten des liegenden
    Menschen. Internationaler Kongreß für Luft- und Raumfahrtmedizin, München, pp 8–9
Vogt LH, Mertens H, Krause HE (1978) Model of the supine human body and its reactions to
    external forces. Aviat Space Environ Med 41 (1):270–278
Vogt LH et al. (1979) Head movements induced by vertical vibrations. AGARD, Lissabon,
    B 11-1-13
Wagner R, Zerlett G (1982) Berufskrankheiten der Berufskrankheiten-Verordnung (BeKV)
    6th edn. Kohlhammer, Stuttgart
Weiner U (1973) Eletromyographische Untersuchungen am sitzenden Menschen während Be-
    lastung durch horizontale Sinusschwingungen. Dissertation, University of Mainz, pp 1–70
Weis EB, Primiano FR (1966) The motion of the human center of mass and its relationship to
    mechanical impedance. Hum Factors 399–405
Whedon GD (1949) Modification of the effects of immobilization upon metabolic and physio-
    logic functions of normal men by the use of an oscillating bed. Am J Med, pp 684–710
White GH, Lange KO, Coermann RR (1963) The effects of simulated buffeting on the in-
    ternal pressure of man. Human Vibration Research, Pergamon, New York, pp 1–27

Whitham EM, Griffin MJ (1978) The effects of vibration frequency and direction on the location of areas of discomfort caused by whole-body vibration. Appl Ergon 9 (4):231–239

Wickström G (1978) Effect of work on degenerative back disease. Scand J Work Environ Health 4:1–12

Winkelholz EA (1967) Messung der Schwingungseigenschaften des Systems Fahrzeugsitz – Mensch. VDI Ber 113:109–112

Witt AN, Fischer V (1980) Vibrationsbedingte Wirbelsäulenschäden bei Hubschrauberpiloten. Research report Wehrmed, BMVg-FBWM 80-2, 1–118

Yamazaki K (1977) The effect of whole-body vibration on human performance and physiological functions. Part 1: Experimental study of the effect of vertical vibration on human sleep. Ind Health 15:13–21

Yokohama T, Osako S, Yamamoto K (1974) Temporary threshold shifts produced by exposure to vibration, noise and vibration-plus-noise. Acta Otolaryngol 78:207–212

Yonekawa Y (1975) An exploratory study of the evaluation of repeated shocks. Human response to vibration. Proc Inst Sound Vibr, Southampton, 65

Yonekawa Y (1978) Evaluation of whole-body transient vibration by finger-tip plethysmogram. Ind Health 16:55–71

Yonekawa T, Miwa T (1972) Sensational responses of sinusoidal whole-body vibrations with ultra-low frequencies. Ind Health 10:63–76

Yules RB (1967) Motion sickness and the reticular formation. Arch Otolaryngol 86:131–132

Zagorski J et al. (1976) Studies on the transmission of vibrations in human organism exposed to low-frequency whole-body vibration. Acta Physiol Pol 27 (4):347–354

Zand SJ (1931) A study of airplane and instrument board vibration. SAE-J 2:263

Zand SJ (1932) Vibration of instrument boards and airplane structures. SAE-J 3:445

Zeller W (1949) Maßeinheiten für Schwingungsstärke und Schwingungsempfindungsstärke. Automobiltechn Z 51 (4):95–97

Zerlett G (1963) Eigene Beobachtung (Berufskrankheiten-Anzeige)

Zerlett G (1982) Beurteilung der Auswirkung von Ganzkörperschwingungsbelastung auf den Menschen. VDI Ber 45:35–42

Ziegenrücker GH, Magid EB (1959) Short time human tolerance to sinusoidal vibrations. WADC Technical Report, Wright Patterson AFB, Ohio, pp 59–391

Zimmermann G (1966) Gesundheitliche Schädigungen bei Traktorfahrern mit besonderer Berücksichtigung der Wirbelsäule. Presented at the 15th international congress on occupational medicine, Vienna

Zorn E (1976) Lärm und Vibration als schiffahrtsmedizinisches Problem. Congress on occupational health, Budapest

12. Medical Terminology

Abdomen	Portion of the body between the thorax and pelvis
Adynamia	Lack or loss of the normal or vital powers; asthenia or weakness
Amotio retinae	Retinal detachment; inner layers of the retina are separated from the pigment epithelium
Aorta	Main trunk from which the systemic arterial system proceeds
Appendicitis	Inflammation of the vermiform appendix
Arteriole	Minute arterial branch
Arthrosis	Degeneration in a joint
Arthrosis deformans	Degenerative process in the joint
Aseptic necrosis	Localized, nonseptic death of tissue or organ
Atrium	One of the two upper cavities of the heart
Baastrup phenomenon	Joint-forming arthrosis of the spinal processes of the lumbar spinal column
Bulbus oculi	Bulb or globe of the eye (eyeball)
Catecholamines	Group of pharmacological substances with sympathomimetic potency
Caudal	Towards the cauda or tail
Caudocranial	From the cauda towards the cranium
Cervical spine	Spinal area in the neck
Cervical syndrome	Condition caused by irritation or compression of the cervical nerve roots, marked by pain in the neck radiating into the shoulder, arm, or forearm
Clay-shoveller's fracture	Avulsion fracture of a spinal process in the lower cervical or upper thoracic spine as a result of an extraordinarily heavy workload; found in workers who shovel clay
Colon	Part of the large intestine that extends from the cecum to the rectum
Cornea	Transparent structure forming the anterior part of the fibrous tunic of the eye
Cortical	Pertaining to a cortex or bark
Cranial	Towards the cranium
Curved back	Kyphosis
Diastole	Phase of the dilatation of the heart, especially the ventricles
Disk hernia	Herniation of an intervertebral disk
Diverticulitis	Inflammation of a diverticulum or circumscribed pouch or sac of variable size, occurring normally or created by a defect
Dorsoventral	From the back towards the abdomen
Dropped stomach	Downward displacement of the stomach (gastroptosis)

Duodenal ulcer	Peptic ulcer situated in the duodenum
Duodenitis	Inflammation of the duodenal mucosa
Duodenum	First or proximal portion of the small intestine
Electrocardiography	Recording of the electrical activity of the heart muscle
Electroencephalography	Recording of the electric currents of the brain
Electromyography	Recording of the electric properties of the skeletal muscle
Epidemiology	Study of the relationships of the various factors determining the frequency and distribution of diseases in a human population
Epigastrium	Region of the abdomen located directly over the stomach
Epiphysitis	Inflammation of an epiphysis (end of a long bone) or of the cartilage that separates it from the main bone
Etiology	Study of the factors that cause disease
Fatigue fracture	Fracture attributed to prolonged high strain
Fibrositis	Inflammatory hyperplasia of the soft tissue (muscular rheumatism)
Flat back	Extended posture of the spine
Fracture	Break or rupture in a bone
Gastric ulcer	Ulcer of the gastric mucosa
Gastritis	Inflammation of the stomach
Gastroenteritis	Inflammation of the stomach and intestines
Gastroptosis	Downward displacement of the stomach (dropped stomach)
Habituation	Gradual adaptation to a stimulus or to an environment
Hemostasis	Arrest of bleeding, either by the physiological properties of vasoconstriction and coagulation or by surgical means
Histological	Pertaining to histology, or that department of anatomy that deals with the minute structure, composition, and function of the tissues
Hypersecretory gastritis	Gastritis caused by hyperchlorhydria
Hypertonia	In general, high blood pressure; excessive tone, i.e., of the skeletal muscles
Hyperventilation	Abnormally increased and deep breathing, resulting in reduction of carbon dioxide tension
Hypoglycemia	Abnormally diminished content of glucose in the blood
Hypoplasia	Incomplete or underdevelopment of an organ or tissue
Intestinal atonia (atony)	Decreased tone of the intestine
Intestinal motility	Natural spontaneous movement of the intestines
Intervertebral foramina	Passage formed by the inferior and superior notches on the pedicles of adjacent vertebrae
Intervertebral osteochondrosis	Degenerative disease of the cartilage and bone tissues of the spine
Intraocular	Within the eye
Juvenile kyphosis	Abnormally increased convexity in the curvature of the thoracic spine in adolescents (Scheuermann's disease)
Kinetosis	Disorder caused by unaccustomed motion (motion sickness)
Kyphosis	Abnormally increased convexity in the curvature of the thoracic spine
Kyphoscoliosis	Backward and lateral curvature of the spinal column
Landolt rings	Rings with a section missing that are used to test visual acuity

Lateral	Denoting a position farther from the median plane or midline of the body
Leptosomatic condition	Having a slight, thin body
Lesion	Any pathological or traumatic discontinuity of tissue or loss of function
Lumbago	Pain in the lumbar region
Lumbar spine	Part of the spine between the thorax and pelvis
Lumbar syndrome	Generic term for all degenerative diseases of the lumbar spine with acute or chronic pain
Lumbosacralization	Assimilation of the lower lumbar vertebrae with the sacrum
Mechanoreceptor	Receptor that is excited by mechanical pressures or distortions
Meissner's corpuscles	Tactile corpuscles
Merkel's corpuscles	Tactile corpuscles
M. erector spinae	Erector muscle of the spine
M. trapezius	Trapezius muscle of the back
Morphology	Science of the forms and structure of organisms
Olfactory system	System pertaining to the sense of smell
Orbita	Orbit; bony cavity that contains the eyeball and its muscles, vessels, and nerves
Os lunatum	Lunate bone in the wrist
Os scaphoideum	Scaphoid bone in the wrist
Osteoarthrosis	Chronic degenerative process in the joint
Osteochondrosis intervertebralis	See: intervertebral osteochondrosis
Paravertebral	Beside the vertebral column
Patellar reflex	Quadriceps jerk (muscle in the thigh)
Peptic ulcer	Ulceration of the mucous membrane of the esophagus, stomach, or duodenum
Peristalsis	Wormlike movement by which the alimentary canal or other tubular organs propel their contents
Plethysmography	Recording of the variations in the volume of an organ, part, or limb and in the amount of blood passing through it
Postmortem	After death
Prevalence	Total number of cases of a disease in existence at a certain time in a designated area
Prostatitis	Inflammation of the prostate
Pseudospondylolisthesis	Forward displacement of a lumbar vertebra
Receptor	A specific molecule on the surface or within the cytoplasm of a cell that receives a specific stimulus
Resection	Excision of a portion of an organ
Respiratory minute volume	The quantity of gas expelled from the lungs per minute
Retina	The light-sensitive internal coat of the eyeball
Retrolisthesis	Backward displacement of one vertebra over another
Roentgenkymography	A technique of graphically recording the movements of an organ or structure on a single X-ray film
Roentgenocinema-tography	Moving picture roentgenography, or making a record of internal structures of the body by passage of X-rays through the body to act on special film

Sacral	Pertaining to the sacrum
Sacroiliac	Pertaining to the sacrum and ilium
Sacrum	The triangular bone just below the lumbar vertebrae
Schmorl's node	An irregular or hemispherical bone defect in the upper or lower plate of the vertebral body
Sciatica	Syndrome characterized by pain radiating from the back into the buttocks and into the lower extremity; pain anywhere along the course of the sciatic nerve
Scoliosis	Considerable lateral deviation in the normally straight vertical line of the spine
Spina bifida	Congenital anomaly characterized by defective closure of the bony encasement of the spinal cord, through which the spinal cord and meninges may or may not protrude
Spondylolisthesis	Forward displacement of a lumbar vertebra caused by spondylolysis
Spondylosis deformans	Degenerative changes of the intervertebral disk with new bone formations at the periphery of the disk
Spongiosa	Inner thin tissue of the bone
Subluxation	Incomplete or partial dislocation
Synapse	Site of functional apposition between neurons, at which an impulse is transmitted from one neuron to another by either electrical or chemical means
Systole	Phase of the contraction of the heart, especially the ventricles
Tachistoscopic tests	Method of testing promptness of visual perception
Thoracic spine	Spine between the neck vertebrae and lumbar vertebrae
Thorax	The chest cavity
Tonus	Tension, e.g., of the muscle
Traumatic	Pertaining to trauma, injury, or damage
TTS	Temporary threshold shift of hearing after exposure to high noise level
Tympanic membrane	Membrane separating the middle from the external ear
Uncovertebral	Pertaining to the joint between the uncinate process of the upper plate of the vertebral body and the lower plate of the following higher vertebral body
Vagotomy	Interruption of the impulses carried by the vagus nerve (first performed by surgical methods)
Vasodilatation	Dilatation of a vessel, especially dilatation of arterioles leading to increased blood flow to a part
Vasomotorium	Vasomotor system of the body, i.e., the part of the nervous system that controls the caliber of the blood vessels
Vegetative	Pertaining to the vegetative or autonomic nervous system
Ventilation	Process of the exchange of air between the lungs and the ambient air
Ventral	Towards the belly
Vessel atonia (atony)	Lack of normal tone of the vascular wall
Vestibulo-ocular	Pertaining to the vestibular and oculomotor nerves or to the maintenance of visual stability during head movements
Vibration-induced vasospastic disease	White-finger disease
Visual acuity	Clarity of vision

Visual cortex	Area of the occipital lobe of the cerebral cortex concerned with vision
Waterfall stomach	An atypical form of hourglass stomach, characterized roentgenologically by a drawing up of the posterior wall; synonymous with "cascade stomach"